RSPB BIRDS

THEIR HIDDEN WORLD

RSPB BIRDS
THEIR HIDDEN WORLD

PETER HOLDEN

BLOOMSBURY
LONDON · BERLIN · NEW YORK · SYDNEY

The RSPB speaks out for birds and wildlife, tackling the problems that threaten our environment. Nature is amazing – help us keep it that way.

If you would like to know more about The RSPB, visit the website at *www.rspb.org.uk* or write to: The RSPB, The Lodge, Sandy, Bedfordshire, SG19 2DL; Tel: 01767 680551.

Published 2012 by Bloomsbury Publishing Plc, 50 Bedford Square, London WC1B 3DP

ISBN (print) 978-1-4081-5262-1

A CIP catalogue record for this book is available from the British Library

Commissioning Editor: Nigel Redman
Project Editor: Julie Bailey
Design: Nicki Liddiard, Nimbus Design

Printed in China by C & C Offset Printing Co Ltd

10 9 8 7 6 5 4 3 2 1

Visit *www.acblack.com/naturalhistory* to find out more about our authors and their books. You will find extracts, author interviews and our blog, and you can sign up for newsletters to be the first to hear about our latest releases and special offers.

Contents

Introduction

It was early morning and distinctly chilly. The sun was not yet above the horizon, but later it would be warm, perhaps even hot. I stood on the edge of the Colca Canyon in Peru, not sure what to expect or where to look. All around towered the jagged peaks of the Andes.

I was not alone: cars and buses had been ferrying people from hotels and hostels from many miles away, and we all gathered here on the side of the canyon, as if at a religious ceremony. Young and old were all here, suitably and unsuitably dressed; all milling together – with cameras ready for the action. Talk, in many different languages, was subdued.

As the sun crept above the horizon, and the first rays brushed the sides of the gorge, a condor rose effortlessly in front of us, hanging on the updraft and supported on massive wings – the largest wings of any living bird.

It was close, so close. There were gasps from the crowd; cameras clicked and whirred, but as we watched it steer its way to a nearby crag we almost missed a second and then a third bird silently emerge from below us and rise up and follow in the path of the first condor. One by one, and sometimes two together, more condors rose from the canyon – so close that it seemed possible to reach out to t ouch a wing tip.

We were impressed. People jostled to get the best position and children were pushed to the front. It was a spectacular show, but the watching crowd was of no interest to the birds. Eventually they passed over our heads and disappeared far away among the mountains. The display was over... until tomorrow when another group of tourists would be here to witness the spectacle again.

I tell this story not just because it was one of those special moments when nature has the power to take my breath away, but also because of its impact on the crowd of spectators. Hardly any were people you would immediately recognise as 'birdwatchers', and there were very few binoculars in sight. Everyone from small children to pensioners was here, and I could not guess how many nations were represented.

The Andean Condor has a 3m wingspan, and a larger wing area than any other bird. Wandering Albatrosses have narrower wings, but with a longer span (almost 4m) from wing-tip to wing-tip.

Children can get caught up in the magic of watching birds, but their concentration spans may be short. If their interest is genuine they absorb astonishing amounts of information, and can put many adults to shame with the breadth of their knowledge.

THE POWER OF BIRDS

Birds appeal to both men and women of all ages, in all corners of the world and across all sectors of society. I find that nearly everyone I meet has a bird story to tell: swans on the park lake; Blue Tits in the garden nestbox; a woodpecker on the lawn. The million-strong membership of the RSPB is an impressive testimony to this wide appeal here in the UK.

The appeal of birds no doubt owes much to their powers of flight, which have long embodied the ideals of freedom and escape, while their voices, colours and journeys have proved equally inspiring. From the biblical dove of peace to the beloved Magpies of Newcastle United football club, this appeal has given birds an emblematic role in the art, culture and politics of peoples around the world.

Different people appreciate birds in different ways. Some of

us develop the same emotional attachments to our individual garden birds as those to our household pets. Others are more taken by their aesthetic qualities: the beauty of birdsong in spring or the drama of a flight of geese against the evening sky. For those drawn to a more scientific approach, birds are an endlessly fascinating subject for surveys and research. And for the collectors – those who list or 'twitch' as many different species as possible – watching birds becomes a personal challenge and, sometimes, a lifelong obsession.

In the modern age, this interest is no longer confined to the scientist or expert. Not only are affordable binoculars now widely available but advances in digital technology also allow any enthusiast to take high-quality photographs – or even branch out into video or sound-recording. The internet has fuelled an explosion of information on birds and birdwatching, while social networking sites have generated an online community that shares information and sightings at the

Sir Peter Scott's initial interest in birds was as a sportsman, but he also captured the beauty of wildfowl in his art. Later he became one of the leading conservationists of the 20th century and an inspiration for many people.

Bird tables can add an extra dimension to a garden – bringing colour and movement. If positioned close to a window they can provide enjoyment and inspiration to all ages, and especially to the young and old.

click of a mouse. Today's birdwatcher can use a smartphone to identify a bird, snap a picture and share the sighting with friends worldwide, all in seconds flat.

BIRDS AND QUALITY OF LIFE

Their wide appeal means that birds are excellent ambassadors for all of nature. Through birds, people are drawn into the countryside, look harder at what is around them and become concerned about the erosion of the natural environment on which birds depend.

The coalminer's canary is an apt metaphor of birds' ability to warn us of danger, and illustrates the numerous intangible ways in which birds can benefit humanity. Even governments now recognise that birds can improve our quality of life – something many of us had suspected for years – with research having proved that birdsong can lift the spirits, and that the physical exercise and intellectual stimulation of watching birds can contribute to a healthy lifestyle.

In my years running the junior section of the RSPB I sometimes met children with learning difficulties, who would learn to read because of their love of birds and a desire to find out more about them. It was also noticeable how often an empathy with birds and nature was found in children who had difficulty fitting in with their peer groups, and how good teachers and youth leaders could often use this interest as an educational and social tool. Many of those children captivated by the world of birds retain their interest for life and, as adults, they are likely to pass the interest on to their own children.

At the other end of life's span I know of retirement homes where feeding birds give untold pleasure to those whose days of walking in the countryside have passed. Garden birds close to a window can bring movement and beauty into the last days of our lives.

BIRDS AND THE ARTS

Artists have long found birds an appealing subject, from the numerous species depicted in Egyptian hieroglyphs to the

Frontiers of science and art

Edward Wilson was a talented artist who was on two expeditions to the Antarctic and eventually died there with Captain Scott. He used his skills to record the landscape, birds and other wildlife seen on the expeditions – especially the Emperor Penguins, which had not previously been observed nesting during the Antarctic winter. A book describing the expedition, *The Worst Journey in the World*, became a popular classic of man's endurance in the face of nature.

feather-perfect illustrations in today's field guides. Edward Wilson, doctor on Captain Scott's fateful journey to the South Pole, amassed a huge collection of paintings and drawings, made under gruelling Antarctic conditions, Archibald Thorburn caught the appeal of the gamebirds for his wealthy sporting patrons, and Sir Peter Scott turned from hunter to painter and finally to the leading conservationist of the 20th century.

In 2007 the magic of the dawn chorus inspired British artist Marcus Coates to produce his amazing film Dawn Chorus for which he taught humans how to mimic birdsong, by copying slowed-down bird recordings, filmed them, then sped up the footage to avian pitch again to create a monumental tribute to the natural world.

It is not only in the visual arts that our fascination with birds has found expression. Composers and musicians have long been inspired by birdsong, from the Cuckoo in Beethoven's *Pastoral Symphony* to the Blackbird in the 1968 Beatles' song of that name. Birds in popular music generally symbolize freedom and positivity, as in Bob Marley's *Three Little Birds* (1977) or Nelly Furtado's *I'm like a bird* (2001). The Guillemots, The Eagles and the Housemartins are among numerous bands that have drawn their names from birds.

Literature is also liberally sprinkled with birds. Sixty-four species, from choughs to woodcocks, appear in the collected plays of Shakespeare, while more recent examples include Jonathan Franzen's acclaimed 2010 novel *Freedom*, in

John Clare, the Northamptonshire poet (1793–1864), was arguably at his best when writing about birds. We can now recognise that his charming detailed observations were remarkably accurate, providing a picture of the English countryside at that time. His poem *The Pettichap's Nest* presumably refers to the Chiffchaff.

THE PETTICHAP'S NEST *(extract)*

Built like an oven, through a little hole,
Scarcely admitting e'en two fingers in,
Hard to discern, the birds snug entrance win.
'Tis lined with feathers warm as silken stole,
Softer than seats of down for painless ease,
And full of eggs scarce bigger even than peas!
Here's one most delicate, with spots as small
As dust, and of a faint and pinky red.
—Well! let them be, and Safety guard them well;
For Fear's rude paths around are thickly spread,
And they are left to many dangerous ways.
—Stop! here's the bird—that woodman at the gap
Frightened him from the hedge:—'tis olive-green.
Well! I declare it is the Pettichap!
Not bigger than the wren, and seldom seen.
I've often found her nest in chance's way,
When I in pathless woods did idly roam;
But never did I dream until to-day
A spot like this would be her chosen home.

which a campaign to save the threatened Cerulean Warbler forms a pivotal subplot. Numerous poets have found similar inspiration, from Gerard Manley Hopkins' *Windhover* (1877) to Ted Hughes' *Thrushes* (1960), often using birds to evoke time and place, or to link the enduring power of nature to the mortality of man. Children's literature that harnesses the appeal of birds ranges from Arthur Ransome's classic *Great Northern* (1947), about children protecting Britain's first ever nesting Great Northern Divers from unscrupulous egg collectors, to Hedwig and the other owls that provide a postal service for wizards in J.K. Rowling's best-selling *Harry Potter* series (1997–2007).

In cinema, Hitchcock's thriller *The Birds* (1960) and Ken Loach's gritty social commentary *Kes* (1969) are among two classic movies that illustrate how the appeal of birds transcends all genres on film. Cartoon birds from Roadrunner to Woody Woodpecker have long kept children entertained, while *Happy Feet* (Warner Bros), a 2006 animated movie about penguins, demonstrates how this entertainment can also deliver a powerful environmental message.

RICH IN FOLK HISTORY

There is a wealth of folklore, stories and even nursery rhymes that can be traced back to the early culture of Britain and western Europe. Obviously this is not limited to birds alone, but it was a bird, a Robin, that covered the 'babes in the wood' with leaves. Wrens were linked with the renewal of the seasons, with winter wren hunts taking place in parts of Britain. Ravens were associated with death – with good reason, as they would have been scavengers on many of the fields of battle. Owls seem to have had a duel identity, being thought both wise and an omen of death.

A snapshot of our long-standing affection for birds can be seen in some of their old country names. Jenny or Kitty were both common names for the Wren. Others included Polly Washdish for Pied Wagtail, Tom Tit for Blue Tit and Jack in-a-bottle for Long-tailed Tit. Mavis could be a Song or Mistle

The Book of St Albans, published in 1486, states which raptors are suitable for differing ranks of social hierarchy, starting with an eagle for an Emperor and a Gyr Falcon for a King and going down through the ranks with a Peregrine for a Prince, a Merlin for a Lady, a Hobby for a young man, a Sparrowhawk for a priest and a Kestrel for a knave or servant. Mews for keeping birds of prey were built in castles and skilled servants were employed to look after them.

Thrush – and, of course, Mavis was also adopted as a girl's name.

Many birds feature by name in the Bible. Noah sent out both a Raven and a dove from his ark. The Robin's red breast supposedly originated from the bloody thorns from Christ's crown, and St Frances and St Cuthbert are both associated with feeding birds.

In other cultures and religions many birds have acquired totemic status. The ancient Egyptians regarded the Sacred Ibis as the incarnation of the god Thoth, while to the Maya and Aztec Indians of Central America the Resplendent Quetzal was emblematic of the deity Quetzalcoatl. Even today, in the tiny African kingdom of Swaziland, the feathers of the Purple-crested Turaco are worn by royalty and symbolic of divine kingship.

BIRDS FOR FOOD AND SPORT

I once stood on a clifftop in northern Norway and heard a local guide telling tourists about annual festivals to welcome back seabirds. The audience was impressed: "How quaint," I heard someone say. Quaint, yes, but this was about survival – the seabirds would have been migrants returning from the ocean, and local people would welcome them because the birds, and especially their eggs and young, would have been a supply of summer food. The arrival and departure of migrants must have been well known since the earliest times, when the birds became an essential component of the human diet.

Hunting birds must have begun as soon as humans learnt to throw stones or spears – or perhaps even earlier – but hunting with birds of prey reached its peak of popularity in medieval times with the sport of falconry. This probably began in the Far East around 4,000 BC, and became popular in Europe and the Middle East. It was further developed in Europe after the Crusades. Social prestige went hand in hand with the birds chosen to hunt.

A HIDDEN WORLD

Most people I meet have, like me, no formal scientific education, but they are curious about wildlife – and especially birds. Yet so many are unaware of the real-life natural dramas that unfold around us; the world of birds just beyond our doors. The more one looks into nature the more one finds – and the deeper the mysteries become.

This book is my humble attempt to share some of the knowledge and stories I have discovered on my journey through a life with birds. Some are my own observations, much of the substance has been gained from colleagues and friends, and other information comes from a massive library of books and papers I have accumulated. It is arranged in chapters; following the lives of birds from egg to death.

My examples are drawn mostly from Britain and Europe, but I have sometimes included examples from other regions to illustrate even greater diversity.

Never was there a better time to start looking harder and asking questions. Gradually scientists are discovering new and fascinating information, and this book shares many new insights into bird behaviour so that every walk in the countryside and every observation of wild birds visiting a garden can be put into context. In this way, I hope that something of the wonder of our avian neighbours will be revealed if only we take time to stop and look.

Photo overleaf:
Black Grouse display.

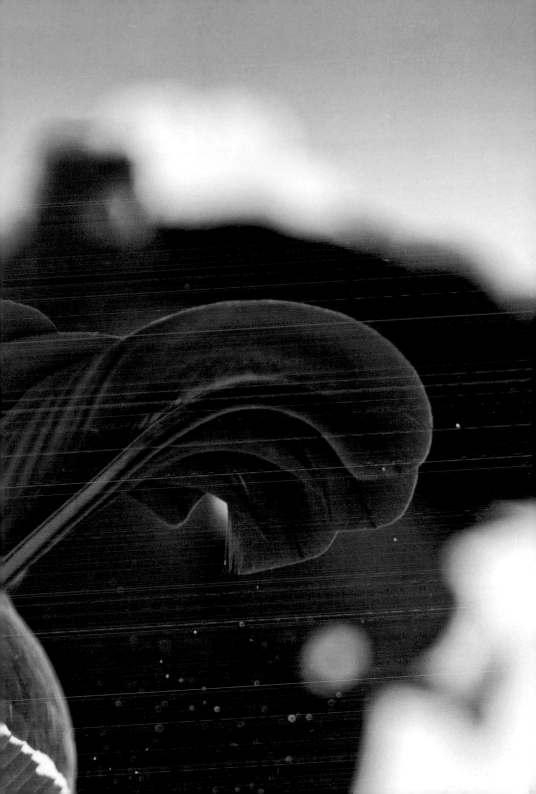

All about eggs

Producing eggs is one of the evolutionary links between birds and their prehistoric ancestors, the dinosaurs. Tending and protecting eggs can be a dangerous and time-consuming activity, but it allows the young to develop outside the female's body and thus does not limit her ability to fly.

WHAT IS AN EGG?

A bird's egg is a complex structure. It is formed in a female bird's oviduct, and contains the nutritious yellow yolk (food for the developing embryo), the springy transparent albumen, which provides a protective suspension system for the developing chick, and two tough membranes holding it all together. All this is contained within a strong outer shell, which is about 94% calcium carbonate.

Eggs that will hatch naked, inactive chicks tend to be proportionally smaller and have a content that is about 20% yolk. Chicks that are fluff-covered and mobile soon after hatching (for example Lapwings) come from larger eggs that are 35% yolk, up to a maximum of 70% in the exceptional

Bird eggs come in an amazing array of colours and patterns.

'Clutch' is the name used for all the eggs laid and incubated together in a single nest. Incubation is the action of 'sitting' on the eggs to ensure they remain at a constant warm temperature, enabling the embryo to mature and eventually hatch. Many birds moult some breast feathers and develop a bald 'brood patch' on their undersides which sits against the eggs during incubation. There are many blood vessels near the skin surface, which helps to ensure the temperature of the eggs reaches that of the parent's body.

Australian megapodes. Although the size, shape and colour of eggs varies between species, within a species it is remarkably consistent – although there are exceptions as we shall see below.

SIZE

The sizes of eggs naturally vary with the size of the bird that lays them, with the smallest eggs in Britain and Ireland being laid by the Goldcrest, about 10mm long, and the largest by the Whooper Swan, at 126mm long. The fastest to hatch may be those of Lesser Whitethroat, needing only 10 days of incubation, while seabirds are among the slowest developers, with Fulmar eggs taking 52 days. Wandering Albatross parents, in the Southern Hemisphere, must spend nearly three months incubating their single egg – it's no surprise this species breeds only once every two years.

The eggs in a single clutch are approximately the same size, but there is a tendency for the last egg laid by raptors and other birds with variable food supplies to be slightly smaller

Research shows that a complete
brood of four or five young
Chaffinches survive better than
when only two or three hatch.
The reason appears to be that a
complete 'rosette' of small young
are better at supporting each
other, and food gets shared out
more evenly. As a result more
young survive.

This means it produces a smaller chick, which may not survive
in years when food is hard to find, as it will not be able to
compete with its larger siblings for food.

The relative weight of the egg also varies between species,
and it seems – not surprisingly – that the more aerial species,
such as Swallows, lay the lightest eggs.

NUMBER

The number of eggs laid in a clutch, and the number of clutches
per year, also varies from species to species. Many long-lived
seabirds, like the Gannet, only lay one egg and only once a year.
The much smaller and relatively short-lived Blue Tit lays one
clutch a year, but it may contain 10 or even 12 eggs. Blackbirds
produce three to five eggs in a single clutch, and many have
two or even three broods in a year. Some species, such as
Robins, produce more broods in southern parts of their range,
taking advantage of a longer season of good food availability.

The record for the number of eggs per clutch laid in the |
UK belongs to the Grey Partridge. It generally lays 13–16 eggs
but clutches of over 20 have also been found to be laid by the
same female.

SHAPE

We all know what the term 'egg-shaped' means, but the shape of a typical egg is properly called pyriform or 'top-shaped'. The shape varies slightly from species to species. Some hole-nesters, such as Kingfishers, lay almost spherical eggs, while Guillemots on cliff ledges lay very top-heavy, pointed eggs. In streamlined species like the Swift, the eggs are quite long and narrow.

COLOUR AND PATTERNS

The colour of eggshells varies from pure white in owls, to a beautiful sky blue in Dunnock, and speckled in Blackbirds. Some patterned eggs are very distinctive; Yellowhammer eggs, for example, have irregular dark lines on the shells, giving the species its old country name of 'Scribbling Lark'.

Ground-nesting birds, such as gamebirds and waders, tend to have wonderfully camouflaged eggs, making them difficult for predators to spot when the adults leave their nests. By contrast, hole-nesting birds have no need for this camouflage and many lay white eggs – indeed, it may help the bird to settle safely onto its nest when the eggs show up clearly in the dark.

The Great Crested Grebe nests in the open, but covers up its plain-coloured eggs with water weed when it leaves the

Guillemots nest on exposed cliffs and build no nest. A single egg is incubated on their webbed feet. Changeover between parents is precarious, but it helps that the egg is long and narrow with a wide bulbous end, with the weight of the embryo generally at the blunt end, helping the egg to spin rather than roll if knocked. The chief danger is the egg getting stuck in a crevice in the rock.

Buntings build open nests so need camouflaged eggs. Those of the Yellowhammer bear chaotic dark 'scribbled' markings giving rise to the old local names of Scribbling or Writing Lark.

EGG COLLECTING

The beauty and variety of birds' eggs made them very attractive to collectors, especially in Victorian England. Vast collections were amassed and a scientific study was established: oology, the study of eggs. It was from these collectors that much early information about eggs and nests was gathered, but at the expense of the birds. Nests of rare, sought-after species were regularly plundered to meet the ever growing demands of disreputable collectors who, in some cases, employed local people to collect the eggs for them. For example, in 1852 a Yorkshire collector called John

Wolley regularly paid ghillies in Scotland to send him Osprey eggs – when that species was on the verge of extinction in Britain.

Many private collections have been acquired by museums and are now available for public scrutiny. They are still valued for research, and allow the public to appreciate the beauty and variety of eggs without any destruction or disturbance to wild birds.

Egg collecting (or 'egging') used to be a common childhood hobby, especially in the first half of the 20th century. It was not unusual for small boys in particular, including myself, to be encouraged to collect eggs. Nests

containing eggs regularly appeared on 'nature tables' in primary schools. Egging is now rightly illegal to protect the birds, but it did create a generation of people who were knowledgeable about birds' breeding biology. Many reformed eggers went on to become excellent field workers and conservationists.

Early bird protection laws were created to prohibit the collecting of birds' eggs on the Yorkshire cliffs in 1869. But it was not until the Bird Protection Acts of 1954 that egg collecting became illegal, except for a few pest species and eggs taken under licence for research.

nest. This activity usually leads to stained and discoloured eggs. Most species lay similar-looking eggs, but some, like Guillemots, have a wide variety of patterns, which may help individual recognition in a crowded clifftop colony.

EGGS AS FOOD

Obviously, the chicken egg has been an important food item for centuries, but in times gone by eggs of wild British species were also key components of human diets, especially in rural communities. The most extreme example of dependency on wild bird eggs was perhaps on the remote island of St Kilda off the coast of Scotland, where locals harvested seabirds' eggs each year. In many other places, gulls' eggs were regularly taken where these birds were common, and plovers' eggs (presumably Lapwing) appear as an ingredient in many old recipe books. Elsewhere in the world, communities still collect wild birds' eggs, not always at a sustainable level. Sooty Terns in the tropics have been known to relocate their colonies to suboptimal habitats, following intense egg-collecting pressure.

EGGS IN RESEARCH

Old collections of eggs became important in the 1960s and 70s when several bird species started to decline sharply, especially birds of prey – and the Peregrine Falcon in particular. Field observation showed Peregrine eggshells were so thin that they were breaking in the nests. Comparisons with old collections showed this thinning was linked to the arrival of certain chemicals in the environment, especially the insecticide DDT.

The discovery of the cumulative effects of DDT and similar persistent insecticides within food chains led to these toxic substances being banned for general agricultural use. Subsequently, Peregrines and other top predators gradually recovered their numbers. Interestingly, recent research indicates that there has also been a progressive thinning of some eggshells in Britain since the Industrial Revolution in the mid 19th century, which continues today. These new results do not bode well for the future.

When the chicks hatch, the eggshells are usually removed from the nest. Sometimes part of the shell is eaten by an adult – perhaps for its calcium content – but more often it is dropped some distance from the nest. Scattering shells away from the nest prevents a tell-tale pile of shells that might give away the presence of a nest to potential predators.

Feathering the nest

There was a time, not so very long ago, when children would head out into the countryside in spring, on the hunt for bird's nests. It was a challenge to find different nests, and their eggs became prized possessions. Thankfully, fashions – and the law – have changed. Now that nests are protected, the subject of nesting has largely slipped from public awareness. But out there in the countryside, and beyond our shores, are some wonderful examples of avian architecture. Nests in all their variety demonstrate the adaptations that help birds breed successfully in virtually every kind of habitat on land – from the middle of lakes to the centre of deserts, and from tropical forests to Antarctic wilderness.

Under cover and made of mud pellets with some vegetation, this Swallow's nest has been built well out of reach of predators.

WHAT IS A NEST?

At its simplest, a nest is somewhere to lay eggs and contain the young until they are ready to leave – which may be just a

The Goldcrest is Britain and Europe's smallest breeding bird. Its nest is a delicate cup of moss, lichen, feathers, hair and spiders' webs, just 9cm in diameter. Suspended among twigs near the tip of a conifer branch, it is usually protected from the prevailing wind, and shelters nine to 11 young.

couple of hours after hatching. In many cases, the site selection and nest-building are also important in developing a bond between a pair of birds, and the shape of the nest cup may even help stimulate the female into egg-laying. So important is this aspect of a bird's life that the whole business of nesting is highly ritualised.

On town park lakes in spring, it is possible to see an example of this behaviour. The male Mute Swan pulls up plant material and passes it over his shoulders to the female, and it is she that arranges the nest. The male then guards his territory, the nest and his offspring.

A Mute Swan's nest may be up to 4 metres across the base. At the other extreme are the beautiful tiny cups, barely wider than a 20p coin, made by some hummingbirds. Many large birds of prey, like eagles and Ospreys, use the same nest year on year and build it up a little more each breeding season. Long-established nests can become truly gigantic.

As we shall see, there is no single nest design for all birds. There are birds that build in trees and those that nest on the ground, and many more which build no nest at all.

THE SHAPE OF A NEST

When you picture a 'typical nest', you'll probably imagine a simple cup of twigs with a soft lining; and many birds build just that. A female Linnet builds a cup of grasses, moss, tiny twigs and roots. Into this she lays four to six eggs, which will be incubated for 11–12 days, and the young will be fed in the nest by both parents for a further 11–12 days before they fledge. The nest will not be used again.

Crows are also cup-builders, though theirs are larger and less tidy versions. The Carrion Crow uses quite sizeable sticks, but the construction is not simple, and comprises four layers. To a foundation of short stout twigs it adds a layer of turf and moss, then a layer of smaller twigs, stalks and roots and, finally, a lining of soft materials such as grass and bark.

Woodpigeons and Turtle Doves build nests that are little more than a stick platform, which looks hardly strong enough to support eggs and young. In some cases the structure is so loosely constructed that it is possible to see the eggs by looking up from below!

A combination of moss, lichen and spiders' webs makes for a remarkably strong and also flexible structure, ideal for when the young Long-tailed Tits grow and move about inside. It is lined with feathers – reputedly as many as 1,000!

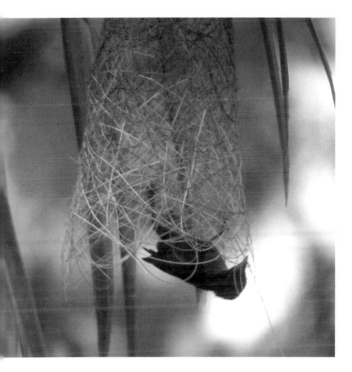

Weaving wonders

Among the 116 species of weavers, sparrow-like birds mostly found in Africa, are some of the most exotic nest-builders. Red-vented Malimbe, a weaver from West Africa, builds a nest from palm fibres in the shape of an inverted sock. The weaving resembles delicate basket-work. The nest chamber is at the top and a long tube hangs down just long enough to prevent snakes from reaching the eggs.

Put a lid on it

Some nests have roofs. One design is a domed nest with an entrance through a hole at the side. Magpies build such a structure – a large and untidy ball of sticks in the branches of large bushes or small trees. A much smaller and more attractive example is made by the Long-tailed Tit. This is a beautiful oval nest of moss and lichen, held together with spiders' webs and lined with numerous small feathers, making a cosy nest for six to eight young.

Hanging nests

Hanging a nest from a branch helps protect it from predators. The Penduline Tit takes its name from its suspended nest, which is domed with a short tube for an entrance. The nest is made of mostly plant down and animal hair, and has the appearance of felt.

Communal living

Social Weavers nest in colonies, in what amounts to one huge, multi-tenanted nest. The first structure to be built is a thick thatch, and under it pairs of weavers make their individual nest chambers – up to 600 of them. The birds nest in semi-desert and benefit from the dense, thatch-like materials. As with a thatched cottage, the 'roof' helps keep the nest chambers cool during the heat of the day, while at night it provides insulation against the cold.

Some species living in other parts of the world have even more dramatic hanging nests. Oropendolas are colourful forest birds found in Central and South America, which build large hanging nests in trees. Many pairs nest together in a colony, and from a distance the groups of pendulous nests can look like hanging fruit.

WHERE TO NEST?

Birds have exploited a wide range of sites for nesting. The site helps determine what the nest will be like and what materials will be needed. Different species have tended to adapt in different ways, but the critical factor is the availability of enough food within a reasonable travelling distance to rear a family.

Out on a limb

Nesting in trees is an obvious way to gain some protection from non-flying predators, but there are endless variations on the theme. Some birds, like the Mistle Thrush, will select a branch fork, or a point where a branch meets the trunk, and build a cup-shaped nest there.

Different parts of a tree are used by different species. Goldfinches and Goldcrests often build among the

outermost branches, Rooks and Carrion Crows near the top
and Woodpigeons make their quite rudimentary nests on
lower branches that are strong enough to support their
growing young.

A bird in the bush

Many small birds choose to nest-build in a bush quite low
to the ground. They rely on the dense (and often thorny)
mesh of stems to keep the nest out of view and reach of
predators. Species that pick sites like this include Song Thrush
and Blackbird. Isolated bushes may be used, but the more
continuous cover offered by a good mature hedge is even
better, and is the reason that you shouldn't cut your hedges
until after the breeding season has finished.

Tree holes

For a nest-site which is not only out of reach but out of sight
to many predators, a tree hole is ideal. Mature trees naturally
develop holes – often where a branch has dropped and the
wood has rotted. Blue Tits and Great Tits nest in small holes,
gathering moss and packing the bottom of the cavity before
building a cup of grass, hair, wool or feathers.

Other small hole users in western Europe are Pied

Pied Flycatchers are at a
disadvantage in woodland
where there is a shortage of
holes, as by the time they migrate
back to the UK the best sites are
occupied by resident species.
Erection of nestboxes in suitable
woods can significantly help
local populations of these
beautiful birds.

Flycatchers and Redstarts – both migrants from Africa that return here in spring. They are in competition with the tits which, being resident, often secure the best sites before the migrants return.

Slightly larger holes are used by Starlings, and even bigger ones by Jackdaws and Little Owls. Barn and Tawny Owls also use large cavities in trees. There can be competition for holes, especially in young woodland, with larger or more aggressive species sometimes ousting others. Tree Sparrows will sometimes build on top of an existing nest of another species, even if it already contains eggs or small young.

Locked in • The Nuthatch's ability to reduce the entrance to its nest with plaster means it can use a wider variety of cavities than some other species and still keep out dangerous intruders. In Africa and Asia, hornbills have taken this precautionary plastering a stage further. Male and female will work together to reduce the size of the hole, and eventually the female is sealed inside with only a narrow slit through which

Getting plastered

Nuthatches use holes in trees and they will fill the cavity with bark flakes. They will also use mud to plaster up a too-large entrance until the hole is small enough to admit them but not bigger animals. The mud sets hard and prevents larger birds from taking over. The instinct to plaster is so strong that when Nuthatches use nestboxes they will fill up any cracks, and even add mud to the roof.

the male can pass food.

In the case of Monteiro's Hornbill, the female is responsible for sealing herself in the nest, using her own droppings as 'cement'. After she breaks out, up to 25 days after the eggs have hatched, the chicks are large enough to reseal their nest hole with their own droppings.

Drill your own · Most woodpeckers excavate their own nest holes by chipping steadily away at dead or living wood to make a suitable chamber (the rapid drumming of woodpeckers that you may hear in spring is a form of communication, and nothing to do with nest construction).

Great Spotted Woodpeckers excavate a chamber 25–35cm deep and 11–12cm wide, with an elliptical entrance of 5–6cm. This substantial hollow is made by both sexes in 14–25 days. Most woodpecker nests are only used once by their makers – but are occupied by other species in future years.

Cavity-nesting and winter meat · Some tree-hole nesters will also utilise other holes. Kestrels, Stock Doves and even Barn Owls will sometimes use cavities in rocky cliffs or walls.

Rock Doves, the ancestors of our feral pigeons, originally nested in cliff crevices and on ledges, and were able to adapt to breed in purpose-made dovecotes. As they breed in most months of the year, their chicks became a dependable supply of fresh meat in winter. It is not unusual to find dovecotes built into the walls of castles for precisely this reason.

Life on the ledge · Several other species which use cliff faces as nest-sites have found the transition from cliff to building quite easy. Kestrels moved into towns and cities during the 20th century, nesting on high ledges and finding food on wasteland, along railway tracks and on roadside verges. Urban Kestrels also tend to catch more small birds than their rodent-eating country cousins.

By the 1990s, Peregrines also started to move into cities in many parts of Britain – using tall buildings as replacements

The Black Redstart is another cliff-nester which has also adapted to nest on buildings. On the continent it can be a common town bird, but in Britain it was mainly a winter visitor and only a sporadic breeder until it moved into and nested in London bombsites and dockyards during the Second World War. It continues to nest in cities, including London, and also on coastal power stations.

The large and historic broch on the island of Mousa in Shetland has a colony of Storm Petrels nesting in the gaps in its stone walls. Like other petrels and shearwaters, these small birds are vulnerable on land and spend most of their lives out at sea. To avoid predators they nest close to the sea and only visit their nests after dark.

for cliffs and feeding on local feral pigeons, or commuting to the local countryside to find more traditional prey. In 2001 they started nesting in London and can now be seen frequently at several sites including perched on the 'chimney' of the Tate Modern on the South Bank.

Burrows and tunnels · Some species nest underground. Kingfishers dig tunnels into riverbanks. They may have two or even three broods in a year; compensating for losses if early broods are swamped with rising water levels. Should a nest be lost the adults are capable of switching back into breeding mode within 24 hours.

Many of our seabirds are tunnellers, on exposed clifftops or islands close to their feeding grounds. Puffins can dig burrows 70–110cm long, but once dug the tunnels may be used for many seasons. Alternatively, Puffins sometimes take over shallow Rabbit burrows.

Second-hand homes · Some birds will reuse the nest of another species – we have seen that woodpecker holes may be reused by other birds. Kestrels do not build their own nests, but choose a nest built by another bird, such as a crow species, in a previous season. Spotted Flycatchers will occasionally use the old nest of another species as a foundation for their own nest.

Most waders nest on the ground in open country, but the Green Sandpiper breaks the mould by frequently nesting in trees near rivers and marshes, and using the old nest of a Song Thrush, Woodpigeon or even an old squirrel's drey. It adds no new material but simply rearranges what's already there for its own use.

Ground level · Not all habitats have trees, cliffs or other elevated nest-sites on offer. Some birds breed north of the treeline, even north of the Arctic Circle, or at higher altitudes. Others nest on lowland heaths, moors, open grassland, marshes, beaches or rocky outcrops. Their only option is to nest on the ground.

There are also a few woodland species that eschew the treetops, like the Wood Warbler, which builds its domed nest of grasses and other plant material on the ground, either among vegetation or under a fallen branch. The adults are aggressive towards small mammals and will hiss and flutter around larger predators to deter them from finding the nest.

Many waders nest in open country, but the Woodcock prefers wet woodland where its wonderful cryptic plumage makes it extremely difficult to spot.

Dig for victory

Puffins are only on land for the short nesting season – for the rest of the year they are out at sea. When digging their nesting burrows, they use their bills to loosen the soil and their feet to push away the debris. They may also incorporate a 'guano pit' or latrine in a short side tunnel.

Larks are ground-nesting birds. All species scrape out a shallow cup and some, like the Bar-tailed Lark of North Africa, add a rim of small stones – piled up like a rampart on one side to give protection from prevailing winds. The stone wall may also assist with thermoregulation.

Our own Skylark is also ground-nesting, and is one of the few songbirds in Britain to nest in completely open sites, away from any trees and bushes. The nest-cup is scraped in the ground, although a natural hollow may be used. The female, accompanied and protected by the male, collects grass with which to line the hollow for three to five eggs.

Some ground nests appear rudimentary, perhaps just a scrape in the ground, although many have some local materials pulled into them. A Stone-curlew's scrape may include small stones, plant remains or even Rabbit droppings. A Ringed Plover on a beach will include pieces of shell, small pebbles and any debris it can find nearby. The result, combined with the camouflaged eggs, is a nest that is very difficult for a human, or a predator, to find.

Watery homes · Some water birds, such as Coot and Moorhen, frequently build out in the water on low branches or among aquatic plants. This helps to protect their nests from non-swimming predators.

Great Crested Grebes sometimes build floating nests of waterweed that are generally anchored to aquatic plants, but may rise with higher water levels. More nest material may also be added if water levels continue to rise.

Fluctuating water levels are a problem for Red-throated and Black-throated Divers. Uncomfortable on land, they nest close to the edges of lochs in Scotland. However, water levels sometimes fall, especially in summer when water is often abstracted for hydroelectricity, and nests may be left high and dry, away from the water margins. Conservationists have overcome this on some lochs by building artificial floating islands that are anchored. The divers have made use of these safe havens, and as a result, breeding success has improved.

With its floating nest, the Great Crested Grebe is often considered the original inspiration for the 'Never bird' of J.M. Barry's Peter Pan.

Branching out

Ospreys add new sticks to their nests each year. Most are picked up from the ground, but some are taken from trees. To secure a branch that's still attached to a tree, the Osprey grabs its chosen branch in flight and allows its momentum to snap off the prize – it is then carried 'torpedo-fashion' to the nest, just as it would carry a fish.

BUILDING MATERIALS

Straw, grasses, roots and twigs are all convenient for small and medium-sized birds to use in their nests. Larger species generally use larger material. Rooks and Carrion Crows use quite big sticks for their tree-top nests. Softer stuff including grass, feathers and wool is incorporated into the nest lining.

Ospreys use even larger branches, and they will sometimes break off small boughs from trees to add to the nest. Young Ospreys in their second year may build a trial nest, even though they will not breed until at least their third year. The first nest built by young Ospreys may be quite flimsy, and may even collapse. However, as the years go by the incumbent pair will continue bolstering the structure, so a typical nest that starts out a mere 50–60cm high may grow, over time, to become 150–200cm.

Insulation and camouflage

Long-tailed Tits and many other small species gather moss and lichen for their nests. These materials are plentiful in the countryside. They give the benefit of insulation, can be moulded into shape, are flexible when there are young in the nest, and help to blend the nest with its surroundings.

Many small nests are held together by cobwebs. Strong but stretchy spiders' silk makes a tough but slightly elastic nest which allows movement, either by the supporting branches or by birds within the nest.

Glorious mud

Mouldable and sticky but drying to a firm crust, mud is a useful building material. Blackbirds include a layer of mud before lining the cup with fine grasses. Song Thrushes build a tidier nest than Blackbirds, and its innermost lining is made of mud, often mixed with dung, rotten wood and leaves. These soft materials are pressed into a smooth cup by the female's breast.

Mud forms the foundations of Swallow and House Martin nests. Swallows build their nests with small mud pellets, mixed with a little plant material to help the mud to bond, and stick these onto a small ledge in a shed, barn or porch. The nest is lined with feathers.

House Martins also use mud pellets to build on the outer

There are reports of House Martins using ready mixed concrete instead of mud for their nests.

Tailor-made

Leaves are often used to line nests, but the tailorbirds, a family of tropical warblers, go even further by contructing the outer shells of their nests from large, living leaves. They use plant fibres as thread and skillfully stitch the leaves together to form a cave-like hollow.

Precious stones

Imagine trying to build a city when there is a global shortage of bricks. For Adelie Penguins, the scarcity of the rocks from which they make their nests turns their breeding colonies into battlegrounds. A careless couple will lose the edges of their nest to thieving neighbours, while a female will seduce males that are not her regular mate in exchange for rocks.

walls of buildings, under eaves or sometimes on cliff faces (where they all nested before man-made stone and brick buildings came along). The mud or clay is collected while very wet and vibrated into place with a shivering movement. It takes on average 14 days to build a new nest from scratch, but repairing an old one from the previous year takes only about three days, and then breeding can begin.

In South America, a group of sprightly songbirds called ovenbirds make huge domed nests of mud. Building may start in the winter, well before the breeding season, when mud or clay is abundant. Several times larger than its owner, the nest can resist rain and is reputed to be strong enough for a man to stand on. Some are built on tree branches that, owing to the weight of the nest, bend and may eventually touch the ground. The nest of the Rufous Hornero is shaped rather like a giant shell, with an entrance that becomes narrower and winds round into an inner nest chamber. Surprisingly, there are observations of juveniles of the previous season helping their parents build a nest.

Bare minimum

Swifts are aerial species; never landing except in emergencies – and then often finding it too difficult to take off again. Swifts originally nested in cavities in cliffs or holes in trees, and in modern times many species use buildings, especially gaps in roofs. What little material they use is what they find blowing in the air. In Asia, swiftlets have taken this theme even further, and use only their own sticky saliva to build their shallow nest cups. The translucent nests are traditionally used to make birds' nest soup.

Natural heat

In Australia the megapodes, hefty turkey-like birds, may never see their own young. They lay their eggs in mounds of sand, stones and vegetation and the eggs are warmed by the heat generated by the rotting material – rather like the heat from a compost heap. Some parents help to regulate the temperature by adding or removing material. The newly hatched chicks dig their way out and go off into the world with no parental care.

No nests

A surprising number of birds do not have nests. There can be a variety of reasons for this, from using someone else's nest to moving the eggs around.

Parasites · The best-known European example of a brood parasite is the Cuckoo, which makes no nest and habitually lays its eggs in nests of other species. There are over 100 species of cuckoos in the world, but only a few are brood parasites without any nest of their own. In southern Europe and North Africa, the larger Great Spotted Cuckoo lays its eggs in crows' nests, especially Magpies.

The Cuckoo is a brood parasite, building no nest and instead relying on foster parents to rear its young.

In other parts of the world are other examples of brood parasites, such as the Black-headed Duck of South America, and the sparrow-sized widowbirds of Africa.

For some birds, laying in other species' nests is an occasional behaviour rather than a way of life. 'Egg-dumping' by one female in the nest of others is quite common, especially among ducks and gamebirds. Some female Starlings have been shown to apparently parasitise their own species, for reasons we don't yet understand.

Wet cement

Nest-building by Wrens appears to be stimulated by rain, when nesting material becomes more flexible. Canaries have been seen dunking material in water, presumably to help make it more pliable.

Rocks, cliffs and ice • While Kittiwakes make seaweed and mud nests on tiny, exposed cliff ledges, other seabirds nesting nearby will use no material. Guillemots breed on ledges and on the tops of rock stacks that emerge directly from the sea. The female lays her single egg directly onto the rock (see page 21). King and Emperor Penguins also make no nest, and as they live on flat areas they can move around (carefully) with their egg balanced on their webbed feet.

HOW TO MAKE A NEST

Each species has its own particular nest design. With only a bill as a tool, birds can create some remarkably complex nests. Some start by dropping material until some lodges and forms a base for a nest. The bill can be used to pull protruding material inwards and push it into the structure, resulting in a rough weave. Some of the weavers even tie rudimentary knots, while others manage to link leaves together. A bird moving within its nest cup smooths out the inner shape, creating a perfect fit.

Ground-nesting species will use their feet to scratch out a nest cup, and use their bodies to mould the shape. The bill is used to draw in and sometimes carry material.

Perfect timing

Most nests are built shortly before the onset of the breeding season. In mild spells in winter, some birds can be fooled into nesting early. Blackbirds, which normally begin to lay eggs in April, have been known to rear young in mild spells in late winter. Blue Tits will start prospecting for potential sites in February, even though their eggs will not be laid until late April or May.

Birds of prey will sometimes frequent their nest-sites in autumn, and some young seabirds, such as Fulmars, will visit potential nest ledges a year or two before they are old enough to breed.

Who builds?

The females usually take the lead in nest building, but we have already seen with Mute Swans it is not unusual for pairs to work together, although it is usually the female that adds the final touches.

Male Wrens will usually build the outer shells of several 'cock nests' – sometimes as many as eight. The female generally

Megastructures

Some tree sites hold enormous nests that are added to year after year, the total weight of a Golden Eagle's nest being estimated at several hundred kilograms.

Most Oystercatchers nest on the ground, but some adopt elevated positions, sometimes on flat roofs. This one has chosen a hollow on a fence post, which is most unusual.

selects one of them and completes the nest by adding the lining – some male Wrens with particularly good territories may attract a second female to use another. The Barred Warbler has a variation on this scenario. As with Wrens the male will build several 'cock nests' for the female to choose from. However if the female disapproves of his choice of sites, she will sometimes find a new site and begin building a completely different nest with the male in attendance.

Reusing a nest

Blackbirds and other species that have multiple broods in a year will often reuse the same nest for subsequent broods. House Martins and some other species, such as Fieldfare, will reuse a previous year's nest. Birds of prey regularly reuse a nest and some eagles' nest-sites are used for many years, not necessarily by the same individuals. Pairs may also rotate between a few favourite sites.

Peregrine Falcons will often nest in an old Raven's nest on a cliff ledge – and Ravens may appropriate sites previously used by Peregrines. In Cheddar Gorge in Somerset, these two species regularly swapped nest-sites for a number of years.

At times other animals use an old birds' nest. Pine Martens, graceful weasel-like animals, have been seen in Golden Eagles' nests, while toads have been found in nests at ground level.

A few birds return to their nest (or a similar nest) to roost. These are generally hole-nesting species. Woodpeckers will roost inside holes in trees. Tits will also use holes but also sometimes sleep in nestboxes in winter.

NESTING ASSOCIATIONS

Noisy neighbours can be annoying, but they may also help scare off more troublesome intruders. The Black-necked Grebe is one of several species that will nest in colonies of other, more aggressive birds, in this case Black-headed Gulls. The gulls help to deter crows, foxes and other predators, making a safer home for the smaller grebes.

Smaller species will sometimes nest in the structure of a

larger species. Tree Sparrows have been observed using cavities in the large twig nests of Grey Herons, while on the continent Spanish Sparrows will 'sublet' spaces in the nests of White Storks.

ADAPTING TO HUMAN NEIGHBOURS

Every so often, birds will be found nesting in strange places, and sometimes these make the local news. A Pied Wagtail pair, for example, nested under the bonnet of an old tractor which was used daily to drive round a farmyard – amazingly the eggs hatched successfully. Jackdaws will nest in chimneys (sometimes with dire consequences for the human occupants if there is no alternative ventilation).

Other curious nest-sites have included letter boxes, clothes-peg bags hanging from washing lines, and cigarette bins outside offices. The culprits are often Robins or Great Tits: probably pairs that have had difficulty finding a more natural hole or cavity.

A helping hand

Changing fashions in architecture, along with more rigorous building maintenance, can dramatically affect local populations of birds which habitually nest in buildings. Swifts are declining in many towns due to a lack of suitable gaps through which they can enter roof-spaces. Birds which nest in tree holes can also suffer when woodlands are managed to remove older trees with natural cavities.

Providing nestboxes helps to redress the balance for these species. The 'standard' rectangular box with a small entrance hole will appeal to Blue and Great Tits, but boxes can be of any size. Larger versions of tit boxes can help sparrows and Starlings. Barn Owls will utilise large triangular boxes, while open-fronted designs may attract Robins and Spotted Flycatchers or Kestrels and Peregrines, depending on size. For Swifts there are special concrete boxes or 'swift bricks' which can be inserted into roofs or even incorporated into new builds.

Blue tits, and also Great Tits, will use almost any cavity in a natural or man-made structure if the entrance hole is the correct size.

Defending a territory

Every spring the woods and fields, moors and heaths, and even our own gardens are partitioned up by invisible boundaries – the divisions between birds' territories. Clues to the owners are there if we take time to listen and look carefully. Birds need territories for breeding and feeding, and staking a claim and defending boundaries is an essential part of their annual routine, but territorial defence takes considerable time and effort.

Once established in its territory a Tawny Owl is unlikely to leave. Young are evicted in the autumn when birds can often be heard calling as they set up new territorial boundaries.

WHAT IS A TERRITORY?

At its simplest, a territory is the area that a bird (or another animal) defends against others, usually of the same species. Territories are generally occupied during the breeding season,

Typical Gannet colonies hold several thousand nests. Nests may be 80cm or less from each other, and both parents need to protect their territory from neighbours, and also from young birds seeking to establish themselves. Fights are frequent, but jabbing, threats and ritualised displays help the resident pair retain ownership.

and include an active nest, but some are defended all year. Other territories are formed in winter to protect feeding sites.

Breeding territories usually include areas where birds will forage, but some species only defend a small area around the nest and feed elsewhere. Gannets, in colonies, have tiny territories around their nests that are pecking distance from their neighbours, but they feed together out at sea. Rooks in a rookery will squabble around their nests, but feed communally in nearby fields.

Home range

For larger birds, the term home range is used to describe both the breeding area and the hunting area. For Golden Eagles, the home range is usually very large – 70 square kilometres or more. The better the area is for hunting, the smaller the home range will be. Most large ranges are not used equally in all seasons, with birds concentrating their efforts where prey is most plentiful. The borders of ranges are defended. aerial displays are usually sufficient to reinforce boundaries between

The home range of a Golden Eagle can cover more than 70 square kilometres, although different parts will be used for hunting at different times of year.

breeding pairs, but young birds that enter a territory will be driven away – the attacks being fiercest in the autumn and when resident adults are preparing for nesting.

Garden territories

While eagles need massive territories or home ranges, Robins are near the other end of the spectrum. Their territories may be little larger than a suburban garden and up to about a hectare (or the size of a football pitch). They are occupied all year, and in Britain and Ireland males and females hold separate winter territories. Generally in autumn the female moves away or occupies just part of the summer territory. Many continental Robins leave their territories and move south for the winter.

DEFENDING A TERRITORY

For many species, it is song that attracts a mate and helps define a territory. The song signals to rivals that a territory is occupied. In addition, birds use ritualised displays that have clear meanings and help to deter intruders. If song and display both fail, then the last resort is chase and attack.

Robins have been well studied and are particularly pugnacious. Intruders foraging in an occupied territory may be tolerated, but if they become aggressive or competitive a song-duel will ensue. If the intruder stands his or her ground then the territory owner will posture, puff up its red breast and

Robins use their red breasts to impress intruders. Their aggressive displays have become ritualised, and different poses are struck depending whether the territory owner is above, on a level or below the intruder.

sway from side to side. Chasing may follow, and around 13% of encounters end in a fight, where there is much fluttering face to face and striking with the feet, as one bird tries to pin the other to the ground and aim blows at its opponent's head and eyes. Usually the intruder gives in, but occasionally an encounter ends in the death of one of the protagonists.

Lekking

For some birds, territoriality is expressed briefly but intensely on a communal displaying ground. This behaviour is called lekking, and one of the British species to engage in it is the Ruff. Males spar and fight with each other to prove dominance, in open areas known as leks. Males defend a territory within the lek that may be only 30–60cm in diameter, and use it so much they sometimes wear out the grass! New or younger birds start on the outside of a lek and gradually move towards the centre. Females only visit to mate, and tend to select the dominant, older birds with darker plumage in the centre of the lek. Less dominant, white-feathered, 'satellite' males join a lek

Male Ruffs grow exotic head plumage for the short breeding season. Dark-plumed Ruffs are dominant at the 'lek'. A few small males do not grow plumes and resemble females. These 'female mimics', known as faeders, are tolerated by the dominant males, and indeed are sometimes mounted by them, but the faeders also enjoy a high rate of success with the females.

Pied Wagtails provide a striking example of how complicated birds' lives can be. In spring they defend breeding territories, like many other species. However, outside the breeding season they roost communally, may feed in flocks, but may also establish and defend winter territories in profitable feeding areas such as along riverbanks.

and will occasionally mate successfully while the dark birds are busy chasing rivals!

Staying put

While many birds are great travellers, others are remarkably sedentary. Once a Tawny Owl has established its territory, it is unlikely to leave for the rest of its life. Young Marsh Tits roam the countryside until they find a suitable territory, but once settled are unlikely to leave. They may join roving flocks of tits and other small birds that pass through their territory, but will leave the flock when it reaches the boundary.

Winter territories

A fascinating study of Pied Wagtails in winter showed that while some foraged in flocks, others defended territories along a river where insect food was available. On the cold, short days of winter there was strong competition for food, and wagtails along the river would drive away competitors. If food became more plentiful territorial males would allow in a 'satellite', often a juvenile or female, which would help in defence against further intruders.

FILLING THE GAPS

When a bird dies, a territory becomes available. Various studies have shown that gaps do not remain for long. The deaths of several Blackbirds in a short time in one garden resulted, on each occasion, in a new bird arriving to take each dead bird's place. This demonstrates how there is a surplus population ready to step in at any time.

Research into Red Grouse populations shows that those not occupying territories live in areas that are poorer in resources, but are available to fill gaps that may arise in the territory-holding population. On the margins they are more vulnerable and their survival rate lower. Conversely, dominant Ruffs which spend time fighting and defending their territories from all comers may have shorter lives than the less aggressive satellites on the fringes of the community.

In Stratford-upon-Avon and many other places, herds of non-breeding Mute Swans gather. These are the next generation, which will in due course set up new territories or replace older adults that die. In winter some breeding birds leave their territories for a time and join these herds.

MAPPING TERRITORIES

From 1962 to 2000, the British Trust for Ornithology ran its Common Bird Census. It relied on observers walking through an area and marking where birds were showing territorial or breeding behaviour. These observations were plotted on a map that soon became a visual representation of the way the area was divided into territories. Although this census has been replaced by the Breeding Birds Survey, the idea of mapping is still a practical exercise for anyone curious about their local bird populations.

A census of Goldcrest territories. Each red dot indicates a sighting (the letter referring to one of 11 visits). Clusters of sightings are ringed to indicate possible territory boundaries and from this the number of territories, and breeding pairs, in a locality can be estimated.

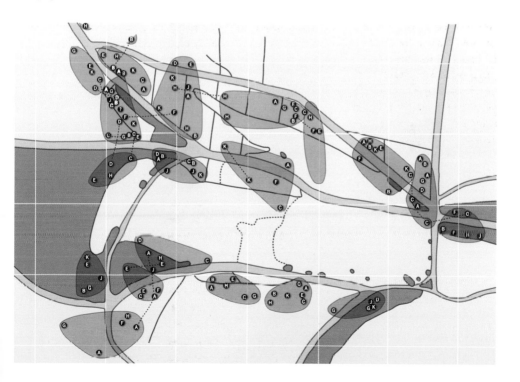

Breeding behaviour

The climax of a bird's year is its breeding season – the annual opportunity to pass on its genes. Nothing could be more important, except perhaps its own survival. All species go to remarkable lengths to ensure the best conditions for rearing their own young in an environment that can, to us, appear hostile – with many predators and fickle weather conditions to contend with. The variety of strategies for success is remarkable.

Young Swallows from previous years will return to the area in which they were reared, but not usually to precisely the same site, although young males return closer to home than young females. This strategy prevents inbreeding.

WHEN TO BREED

Parents need to ensure that their young hatch at a time when there is maximum food available. This needs forward planning, as there is a period of incubation for the eggs that may only take 13–16 days for Blue Tits, but 25–27 days for Grey Herons and a lengthy 52 days for Fulmars. Then there is the time taken to select a mate and build a nest. Research shows older Great

Investing in the future

For Arctic Terns, their short breeding season is the only time they will spend on land. Success depends on a safe site, a good mate and plenty of food. Within around 50 days they lay eggs and rear young. Within around 50 days they lay eggs and rear young, and within weeks of being born young will be independent.

Tits tend to claim larger territories, be more successful at attracting mates and breed earlier – leading to more young successfully fledging.

There are many factors that stimulate birds to begin breeding behaviour. Day-length is critical and so is temperature and rainfall. Sometimes birds can be fooled: Blackbirds have several broods a year and may start nesting in mild spells in February – although very few young from such early broods survive.

For migrants arriving from long distances, the timing of their return to the breeding area is remarkably consistent, often to within a day or two, from year to year. Generally males arrive before females and set up territories. In the case of Swallows, males are very faithful to past breeding sites and so are most females, but slightly less so than males.

Some breeding birds have other requirements. Reed Warblers need a good growth of reed before nesting, and Cuckoos will time their breeding to match that of their hosts. Some Arctic birds may wait for snow and ice to melt or islands to become isolated from ice sheets. Quails migrating from

Great Tits time their nesting so that they can provide their chicks with high-protein food, principally caterpillars. It has been estimated a brood of young Great Tits require 10,000 caterpillars or their equivalent during their nestling phase.

Africa to Europe may stop and breed in southern Europe early in the season before continuing their migration and breeding for a second time further north.

In the tropics, there are wet and dry periods rather than seasons, and these dictate breeding activity. In desert regions, a rainy period at any time of year can stimulate some species to start breeding. Some large species with special requirements and a long maturation period, such as Emperor Penguins and albatrosses, may not even attempt to breed every year.

Some European species are also unpredictable. Crossbills have been recorded breeding in every month of the year, and in Europe it is possible that some will start breeding in August after spruce cones have formed – their season may continue through to the following spring. Crossbills in pine woods may start breeding in December and continue until June.

Synchronised breeding

One mystery is how some birds manage to time their breeding so that it is synchronised with others in the same population. Colonies of Starlings have been found not only to lay their eggs within a few days of each other but also at the same time of day. It is assumed that there is some social stimulation as well as other external factors such as weather that causes this, and it may be a defence against nest parasitism by other Starlings.

Food supplies at egg-laying time may be critical, as females about to lay need sufficient body resources. Geese migrating to the Arctic may have enough stored fat, even after their migration, to get them through egg-laying and the incubation stage of their breeding cycle without much extra feeding.

COURTSHIP DISPLAYS

The approach of the breeding season is characterised by dramatic changes in birds' attitudes to one another. Birds that may have tolerated others of their species or even flocked together in winter now become openly aggressive to one another as territories are established. Courtship displays between birds of the opposite sex become much more frequent.

These courtship displays may be physically demanding, allowing the bird to demonstrate its fitness and general state of health to a potential mate. Blue Tits flutter their wings and have a short gliding display flight. Dunnocks flick one open wing above their backs as they pause, briefly, in their pre-breeding chases. Goldfinches have a pivoting display in which they resemble a mechanical toy as they crouch on a branch and swing from side to side while singing. This display, by both male and female, is thought to strengthen the pair bond as well as having territorial significance to other Goldfinches.

Great Crested Grebes display by expanding the frill around their faces and the crest of black 'tippets'. Often they swim together, shaking their heads and almost fencing with their bills. They have other displays, including the wonderful 'weed dance' which follows a dive, when both birds of a pair bring up waterweed and surface together, treading water and rearing up with their breasts pushing together. They appear to try to present their weed to each other before, eventually, discarding it. This part of the display could represent a ritualised demonstration of food-finding ability – an essential skill for a would-be parent.

Many seabirds are particularly demonstrative, especially those nesting in colonies. Terns, for example, engage in both aerial and ground displays.

Tree Sparrows are surprisingly aggressive when nesting and will sometimes build their nest on top of the nest of another species – even if it contains eggs!

The 'weed dance' is highly ritualised courtship behaviour. Great Crested Grebes dive, gather weeds under water, swim together and rise up as they 'tread water'. Wonderful synchronised behaviour.

Fishing for compliments

On the ground the male tern may carry a fish as he bows and struts with wings bent and drooping. Fish are also used in courtship feeding, which is often a preliminary to copulation. However, in a busy colony females will sometimes receive fish from, and mate with, other males.

The male Common Tern generally carries a fish cross-wise in his bill. In flight a pair spiral upwards, wings beating rapidly, with the female following the male. At the top of the spiral the female passes the male with a curious and exaggerated gesture, with head and neck stretched forwards and wings arched, and makes a begging call. They then slideslip and dive dramatically, close together, flying in parallel. At the bottom of the dive they often swoop up again and repeat the performance. The whole thing is both beautiful and breathtaking.

Gannets have ritualised greeting ceremonies when they return to their nests after an absence of two hours or more. This involves 'fencing' with their dagger-like bills, sparring, a lot of head waggling and each grabbing the other's open bill.

Courtship feeding

Male birds of some species will feed their mates during and after courtship. You may see Robins engaging in this courtship feeding in the garden. It begins with the male foraging and the female begging to be fed – adopting the posture of a juvenile, with wings fluttering. The male then approaches the female and places a food item in her mouth.

Courtship feeding not only helps build up the female's bodily resources prior to egg-laying, but enables the male to demonstrate his ability as a provider, and helps to cement the bond between the pair.

MATING SYSTEMS

Some baby birds require more parental care than others. In general, chicks that are helpless when they hatch, needing to be fed and kept warm, will need the care of two adults over the weeks before they achieve independence. However, for chicks that are down-covered, active and able to feed themselves soon

Courtship feeding gives the female Robin additional food at a critical time, early in the breeding season and just before egg-laying, and it helps to cement the bond between the pair.

Swans are usually faithful to the same mate year after year. Even the migratory Bewick's Swan tends to be remarkably faithful, and when one dies it can take several years for its partner to find a new mate.

after hatching, one supervising parent may be enough. This is mainly what determines whether birds tend to form pair bonds or not, but mating systems can be much more complex.

Monogamy

It makes sense that Golden Eagles, which are long-lived and occupy year-round territories, would form long-term pair bonds. You might expect that migrants like the Osprey, which have to reclaim their territories each year, would be less faithful. In truth, even migrants usually form the same pair as the previous year. However, as the pair migrate and winter separately, this loyalty may be more about familiarity with the nest-site than faithfulness to a particular bird. 'Divorce' is not unknown, but is often linked to failed breeding in the previous year.

Unfaithfulness

Once, monogamy was thought to be commonplace among birds. Where a paired male and female are the true genetic parents of the chicks they are rearing, this behaviour is called 'genetic monogamy'. However, DNA analysis is revealing that behaviour called 'social monogamy' is much more common.

Social monogamy describes relationships where two birds appear to be raising their young together, but where extra-pair copulations are also frequent. This behaviour has been found in common garden species such as Blue Tits and House Sparrows, and we now know that females are frequently attracted to other males breeding in the locality, and will mate with them when the opportunity arises. The result is a brood of chicks with different fathers.

Polygyny

Males of some species frequently form pair bonds with two or more females. This is common with harriers. Polygyny is most common where the population is most dense. Usually it is older and more experienced male Hen Harriers that mate with several females. This arrangement means less time wasted in inter-territory conflicts, but might be a disadvantage in times of food shortage. Some surprising research, however, shows no significant adverse effect, and even a slight increase in young bird survival when four females were nesting close together and supported by a single male.

Male Pheasants often have a harem of females. More mature males with brighter facial wattles are the most attractive to females. Some males attract no partners, and some only one, but harems of up to 18 females have been recorded. Females benefit from feeding during courtship and protection while rearing her family, but the colourful male plays no part in caring for the young.

Polyandry

Females sometimes pair with several males. In North America, female Spotted Sandpipers mate with up to four males and

Dull dad

The Dotterel is a summer migrant to northern Europe that breeds on some of our highest mountains. The female is more colourful than the male, and so it will be no surprise that he incubates the eggs. This leaves the female free to visit another territory or continue her migration northwards, where she may mate with another male and lay a second clutch.

leaves clutches with each of them, although she may assist with the last clutch.

One-parent families

For chicks that feed themselves and live in places where food availability may be limited, or varies from year to year, having two parents can be a disadvantage, because they reduce the amount of food available. One parent to warn of danger and to brood the youngsters at night may be all that is required. Many gamebirds, wildfowl and waders rear their broods solo, and in these species courtship is all about choosing to mate with the bird whose physical condition is best, meaning high-quality genes for the chicks.

The curious case of the Dunnock · In 1853, Reverend Morris wrote a major book on British birds, and cited the Dunnock as a bird with a morally admirable attitude, describing them as "shy, humble and homely ... which many may imitate with advantage to themselves and others". If Morris had known what we know now, he would be turning in his grave!

The female Dunnock takes the lead in territory marking and solicits male attention. About a third of pairings are monogamous, but others involve additional male birds, and after much displaying a hierarchy is established between the males. On rare occasions fierce fights between two equally dominant males may end in the death of one of them – an event I once witnessed on the lawns of the RSPB Headquarters one snowy February day!

With two or more males in attendance, the female will mate with one or all of them. Before mating a male will peck at the female's cloaca and stimulate her to eject any previous sperm before he adds his own.

Sometimes the arrangement becomes even more complex, with a dominant male taking over the territories of several females. Some of these territories may also have attendant males, so a situation arises with complex relationships between multiple males and females.

The benefit of this unusual arrangement (now called polygynandry, meaning males may be polygamous and females polyandrous) is that it ensures the best genes are passed on to the young, and females can benefit from more males assisting with feeding the brood.

BROOD PARASITES

On page 39 we looked at the Cuckoo and its parasitism of other birds' nests, but within-species brood parasitism may sometimes occur. Some female birds will 'dump' their eggs in the nest of another female of their own species. The behaviour is particularly common in ducks, but other species do it as well. For example, research shows that up to 37% of Starling nests may contain eggs that were not laid by the nest-owning female.

Gradually the Starling's complex secrets are emerging. While some pairs are monogamous, other males pair with two or

Dunnocks will flick their wings repeatedly during their courtship. Often only one wing is raised, but sometimes both wings are waved energetically.

Occasionally you may find the blue egg of a Starling intact on the ground, as if it had been laid there. The traditional explanation was that the female responsible had been 'caught short', but research now shows that these eggs originated in a nest and were removed... but by whom?

more females, although without a male's full-time assistance chick survival is usually reduced. Nest parasitism often occurs after first broods have been disturbed and females continue to lay in another Starling's nest.

Some parasitism seems more deliberate; eggs are laid quickly and at a time when the nest owners are away. There is also evidence that some parasitic females remove an egg at the time they lay their own. The resident pair may attempt to remove eggs from nests that have been parasitised. In addition to all this confusion, there have been instances of two female Starlings pairing with the same male and using just one nest.

INFANTICIDE

It's widely known that male Lions taking over a pride will kill the cubs, to bring the lionesses into heat more quickly. A similar unsavoury practice occurs among some bird species, including Swallows. Many pairs are monogamous, but males, especially in colonies, will frequently solicit extra mating with neighbouring females. To counter this, in the run-up to

Male Swallows have longer tail streamers than females and juveniles. More successful male breeders tend to have the longest streamers, and these are the birds preferred by females.

Young Barn Owls in the same brood are of different ages, with the potential for the oldest to eat the youngest if food runs short.

egg-laying male Swallows will shadow their mates to deter advances by other males, and if a male suspects infidelity he solicits extra matings in order to dilute the sperm of the rival.

Sometimes an unmated male will remove all the newly hatched young from a nest and drop them on the ground, where they are doomed. With no young to feed, the female reverts to breeding mode. She may separate from the male that failed to adequately guard her nest and pair with another – possibly the same unattached male that killed her first brood.

SYNCHRONISED HATCHING

Eggs are generally laid at a rate of one a day and clutches vary from one in the case of Fulmars to 15 or more for gamebirds. It benefits most birds if all their young hatch more or less at the same time, so the young are fed together and leave the nest together. This is achieved by delaying incubation until the clutch is complete, resulting in almost synchronised hatching.

A few species, especially birds of prey and owls, have highly variable food supplies and with these species incubation often begins before the clutch is complete, resulting in young of different ages. The youngest and smallest chicks are likely to starve quickly if food supplies run low, meaning the rest of the brood stand a better chance. Dead chicks may be fed to their surviving siblings, and there are cases on record of older Barn Owl chicks killing and eating the smallest chicks in the brood.

Blue and Great Tits frequently use nestboxes in gardens, but research has shown that their clutches tend to be smaller than in woodland nests.

BROOD SIZE

Obviously the size of the brood depends on the number of eggs laid. While this differs from species to species, within a species there is a consistent range: for example a Blackbird typically lays three to five eggs and a Blue Tit eight to 10.

Brood sizes can affect survival, for example a larger number of young, and therefore a greater body mass huddled in a nest, helps to keep individuals warmer and thus saves energy. However, larger broods may also mean proportionally less food is received by each youngster.

External factors can also limit family sizes. Food shortage may reduce the clutch size and the number of young raised, as has been observed in tits breeding in gardens rather than in woodland. Also, clutches laid slightly later in the breeding season and second broods both tend to be smaller.

WHO INCUBATES?

The role of the sexes in incubation is very varied. In Pheasants and ducks it is all done by the females – this being quite an inappropriate activity for a brightly coloured male. Many other species share the responsibility, with the females generally doing most, and especially at night. In just a few species the roles are reversed – for example, in the case of the Red-necked Phalarope and Dotterel it is the male that incubates.

ALTRICIAL OR PRECOCIAL?

On hatching, young birds may be blind and helpless (altricial), like young Swallows that take 18–23 days to develop, grow feathers and become mobile before they leave their nests, or they may be covered with down and able to run about and feed themselves very soon after hatching (precocial), like the chicks of a Pheasant. Some will stay in their nests for a set time, 15–19 days for Wrens, or 90 days in the case of a Gannet. Others, like Lapwing chicks, abandon their nests soon after hatching.

NEST MANAGEMENT

Dealing with a clutch of immobile eggs is one thing. When

those eggs are replaced with noisy, moving chicks, the parents' workload becomes more involved than just keeping those hungry mouths supplied with food.

Egg shells

Broken egg shells present the parents with a problem. If left in the nest they get in the way and may even cover unhatched eggs, so most are removed by the parents. However, dropping them near a nest would be an obvious signal to predators and so they are usually carried some distance away.

Some birds eat part or all of the shells, and parent Flamingos feed the shells to their young. Eating the shell allows some of the calcium content to be absorbed back into the body – which for young birds can help with bone formation.

Droppings

The next problem is the faeces. Nest hygiene is an important activity for most songbirds. After being fed, a chick turns around and hoists its rear upwards, allowing the adult to collect its dropping as it appears. Some droppings are eaten by the parents, but most are covered with a thin membrane which allow the adults to carry them away and dispose of them. Like the shells, they are usually not discarded locally, avoiding an obvious trail to the nest.

Some small ground-nesting birds will move out of their nests to defecate, and young birds of prey have a remarkable ability to squirt liquid droppings out of the nest. Puffin burrows sometimes have a small chamber excavated as a toilet for the young. The prize for cleanliness, however, goes to the lyrebirds of Australia, which take faeces to nearby water and submerge them or, if lacking water, will bury them.

Defending the offspring

Strategies for nest defence vary depending on the site. The best defence is being well hidden or in an inaccessible place. However, many species nest on the ground. A common ploy

Nest sanitation is important, and most species carry away the faeces of their offspring and discard them some way from the nest so as not to provide clues for predators.

Arctic Terns are particularly belligerent and strike humans on the head with the tips of their bills, which is sufficient to sometimes draw blood.

Tawny Owls are particularly aggressive when protecting their young. One famously attacked bird photographer Eric Hosking, resulting in the loss of one of his eyes.

used by plovers and some other ground-nesting species is the so-called 'distraction display'. On the approach of a potential predator, the parent bird will leave its nest and run away trailing a wing, calling loudly and looking distressed and vulnerable, with the hope of luring the trespasser away.

Some ground-nesters are fierce in their nest defence – especially those in colonies. Arctic Terns join forces to dive-bomb trespassers and even strike at them with their pin-sharp bills. Not surprisingly some other species, such as Eiders, often choose to nest among tern colonies for protection.

Some tree-nesters are also very aggressive. Fieldfares swoop at predators, making a lot of noise in the process, and may even defecate on the predator. Press reports of owls terrorising neighbourhoods are usually females protecting their nests.

Fulmars eat fish, offal and other items gathered from the sea. If disturbed at their nest, whether by a gull, skua or rock-climber, they will accurately vomit their half-digested stomach contents at the offender, a smelly and oily mess that can render an intruding bird flightless, and a human very uncomfortable.

Some birds that nest in holes, such as Blue Tit, make snake-like hissing noises if danger threatens, and a few species sometimes carry their young away from danger, either in their bills or, in the case of Woodcock, between their legs.

Food for the family

The biggest responsibility for many parents is to find enough food for the growing family. Most birds that eat seeds as adults will feed their young on high-protein insect food instead. Birds carrying food may be very obvious: a Song Thrush with a bill full of worms, a Guillemot with a catch of fish, or a raptor with prey held firmly in its talons.

More discreet feeders include pigeons, which feed their young on 'pigeon milk', a liquid formed in the crop of both male and female. The youngsters push their soft bills inside the adult's gape and gulp down the 'milk' as the parent brings it up. Other more solid food is also transferred in this way – some seabirds swallow their catch, then regurgitate it either at the chicks' feet or directly into their mouths.

Rooks also disguise their food collecting, but they carry the food for their young in their gullet, in an extendable pouch below the lower mandible. The bulging pouch is obvious on foraging Rooks in spring.

Especially dramatic, a male harrier carrying food will fly into his territory and call the female off the nest. She flies up and slowly circles below the male. At a critical point in this formation flying the male drops his prey, which the female catches in her talons by performing a spectacular barrel-roll.

Collared Doves, like other pigeons, produce 'pigeon milk' in their crops and regurgitate this to their young.

Generally young are fed by their own parents, but juvenile Moorhens can often be seen feeding smaller young from later broods.

Sheltering nestlings

Adults brooding young not only keep them warm, but protect them from the elements. Even when the chicks are larger, the female may shelter them by spreading her wings and shielding them from heavy rain.

Water

Most young birds get enough moisture from their food or in some cases mucus from their parents. Darters from Africa sometimes receive water regurgitated by their parents. However, the most remarkable adaptation is shown by the sandgrouse, which have specially modified feathers for soaking up water which can then be transported up to 30km to provide moisture for their young until they are old enough to fly.

DIVISION OF LABOUR

Once a brood of chicks have left their nest they are highly vulnerable. The young of many species have limited flight, are unwary, and still require feeding for several days or longer after leaving the nest.

Blackbirds, and some other thrushes, split their brood of fledglings between male and female. While their young benefit from more individual adult attention, the survival rate of those accompanying the male can be higher than for those cared for by the female. This situation arises because Blackbirds are multi-brooded, and the female often leaves her fledglings earlier than the male in order to prepare for the next brood.

Outside assistance

Some parents receive extra help when rearing their young. In birds that have multiple broods in a season, chicks from earlier broods sometimes help feed their younger siblings from later broods. This can be observed in Moorhens and House Martins later in the breeding season.

Cooperative breeding may take place when non-breeding birds in a colony are stimulated to feed chicks in the nests of other pairs. Adult Long-tailed Tits that lose their nests may sometimes assist neighbours that are still rearing their young.

The stimulus to feed young is so strong that sometimes a bird will feed the young of another species. I once watched a Wren feeding some noisy young Great Tits in a nestbox. Male Wrens frequently feed other birds' chicks while their own mates are incubating. Young Cuckoos have such effective begging behaviour that they can stimulate other passing birds besides their foster parents to drop food into their mouths.

Anis, from Central America, belong to the cuckoo family but are communal nesters rather than brood parasites. Groups of breeding birds, including some juveniles, defend a territory and up to five females lay their eggs in a single nest. Incubation is shared, although older and more experienced females may have the longest shifts, and once the eggs have hatched the other females share the responsibility for feeding the young.

Guillemots frequently shelter young beneath their wings. These are not always their own, and sometimes a single bird will shelter three or four chicks.

Communal nesting

Colonies of birds are spectacular places to visit. Around the coast of Britain and northwest Europe are some of the world's greatest seabird colonies – many within reach of anyone prepared to take a short boat trip. At the right time of year it is possible to be surrounded by Puffins or looking down on wheeling masses of Kittiwakes. Within a few weeks these birds will have returned to the sea, and the islands and cliffs will be deserted until the following spring.

WHAT IS A COLONY?

It has been estimated that as many as one-eighth of the world species are colonial. However, it is not easy to define a colony, as most 'colonial nesters' also maintain a small territory around their nest, even though they have chosen to nest close to others. Some colonies, such as Kittiwakes, extend continuously along stretches of our coasts and it is far from clear whether this is one large colony or many smaller ones.

Seabirds are great colonial nesters, living close together and as close as possible to their food supplies. However, most will retain a tiny territory around their nests which they will defend from intruders.

On the coast, colonies of Cormorants nest on rocks close to the sea. They will also nest inland, and there the colonies are usually in the branches of tall trees.

Gannets, for example, are highly colonial, but around their nest – pecking distance from their neighbours – is their tiny territory that they defend vigorously. Grey Herons in their tree-top colonies (called heronries) also defend the area around their nests.

Colonies sometimes attract other species to nest amongst them. Black-necked Grebes often nest among colonies of gulls from which they gain protection – the aggressive gulls are effective at driving away potential predators.

Colonies are simply places where many birds come together to breed, and few show any social organization; there is no obvious hierarchy within a colony for example.

The disadvantages of colonial living are clear. There is more competition – for food, mates and nesting spaces. A colony is also more conspicuous to predators than an individual nest, and there is a greater chance of transmission of parasites and disease. There is also a greater chance of the wrong chicks being fed, eggs being dumped in the wrong nest and the mixing of broods. To overcome the disadvantages, there must be powerful benefits for those species that do nest colonially.

THE BENEFITS OF COLONIAL LIVING
In general, birds nest close together because it improves their opportunities to feed, and also nesting in a colony is more efficient than maintaining individual territories.

The open seas provide great feeding potential, but there are limited nesting sites, and seabirds need to nest near to their food. In reality, the best food may sometimes be a great distance away – Guillemots are known to travel up to 200km on a fishing trip, and Gannets are capable of travelling even further – but most food is found much closer to colonies than this, and nesting within easy reach of it is obviously beneficial if not essential.

The far-flung nature of their feeding grounds means that seabirds cannot establish a feeding territory. Therefore all they need is a very small nesting territory, which means they can pack in dense colonies in the best and safest locations. These include sea cliffs and islands where there are suitable ledges and crevices, well protected from most predators.

Rooks and Grey Herons, nesting inland, use traditional sites that are also within easy commuter distance of essential feeding sites.

Birds in colonies can easily form flocks and engage in communal feeding activities. Also, in times of food shortage, birds can follow successful feeders and thus save valuable time searching on their own. In one piece of research, the writers termed colonies 'information centres' in recognition of this important function.

Also, living in a colony with a very small territory reduces the time and effort spent defending the area against other

The size of a heronry appears to be proportional to the amount of food available – with the largest ones in the richest feeding areas.

birds, leaving more time for feeding and rearing young. For inland species, the colony may be in the centre of a food-rich area, saving time in commuting and hunting.

It has been noticed that in many colonies birds breed in synchrony, with eggs being laid more or less at the same time and hatching together. The benefits from this are not entirely clear, but it must help with food finding and predator repulsion if birds are at the same stage of their breeding cycle. Presenting predators with a sudden glut of easy prey in the form of young chicks also has its advantages. This way, more are likely to survive to an age when they can defend themselves than if they hatched one by one over a longer period.

Colonies give protection against predators. While colonies may become obvious targets, approaching danger is more likely to be spotted earlier with more birds on the lookout, and a co-ordinated mobbing attack by many adults is an effective deterrent to all but the most determined predator. Even if some parents are away hunting, others can usually be relied on to drive intruders away.

As we have seen, some colonial nesters are remarkably aggressive. Arctic Terns are particularly courageous in defence of their nests and young – even directing their attacks towards human intruders. A lone bird may not succeed at defending its nest, but en masse the tern colony is formidable.

Sand Martins dig burrows in which to rear their young. Their colonies are sometimes targeted by other hole-nesting birds – especially Tree Sparrows – which take over the nesting chambers for rearing their own young.

IS IT A COLONY?

Think of a colony of birds and one tends to think, perhaps, of herons, terns and Rooks, where their nests are all on public view, but there are many other species that breed close together more discreetly in loose colonies. Starlings are colonial, but it is hardly obvious as their nests are well hidden. Birds from a specific locality leave their small breeding territories and regularly join others from the colony as they search for food in foraging flocks.

The same is true for many, but not all, finches. Chaffinch and Brambling pairs are spread fairly evenly across their breeding habitats, but other related finches, including Goldfinch,

Greenfinch, Linnet, Siskin and the crossbills, are loosely colonial and tend to nest in 'clusters'.

The reason for the difference is their diet. Chaffinch and Brambling are insect-eaters in the summer and their prey is widespread. However, the others are seed-eaters, and are more likely to find their food in dense clumps. Where food is found there is generally plenty of it, so they can afford to feed in flocks – even though they continue to defend a small territory around their nests.

This behaviour helps break the concept of a colony being in a fixed place, because some finch colonies appear to move during the breeding season, with second broods frequently being reared in a different area to the first brood. This move allows the birds to take advantage of new food as it becomes available.

These colonial finches hold individual nesting territories, but they spend most of their time feeding in flocks, even in summer.

WHOSE BABY?

Some ground-nesting birds that nest in colonies have another problem – wandering chicks! Young terns and, famously, young penguins form crèches, with the young birds constantly moving and changing position, which is a challenge for the adults returning with food. It is important that they find their chick and, almost as important, that they don't give the food to the wrong bird!

Mobile young birds can move to the safest parts of their colony, but it is vital that the young recognise their parents and receive their food.

The colonies of this tiny seabird can number thousands, but they remain hidden in crevices, holes and burrows during the hours of daylight.

Penguins and terns use mutual recognition calls, by which juveniles recognise their parents' calls as they return to the colony. This seems tricky at the best of times, but in strong winds and against the background noise of a large colony it presents a real challenge. Experiments to show how this can work have revealed the ornithological equivalent of what psychologists call the 'cocktail party effect', meaning that even over a hubbub of background noise a familiar voice will stand out from the crowd!

THE STRUCTURE OF A COLONY

Some seabird colonies are very old. The Gannet colony on the Bass Rock in the Firth of Forth was first mentioned in literature in 1447.

All colonies are composed of a mix of experienced and inexperienced birds and can be described as either stable or unstable. Stable colonies remain more or less the same size from year to year, while unstable colonies are either increasing or decreasing in size.

In an increasing colony, you are likely to find younger birds nesting around the edges and older, more experienced ones in the centre. However, in a stable colony inexperienced birds will replace pairs lost from anywhere in the colony.

Colonial nests

If we look beyond Europe into Africa we find interesting examples of colonial nesting. As a family, the weavers – a group of small, sparrow-like birds – build some remarkable nests, but none more fantastic than the Social Weaver that we looked at earlier (see page 28). A colony of this species may comprise up to 600 nests built into one huge structure. This is colonial living on a grand scale.

Underground colonies

Some colonial nesters breed underground. They include some of the smaller seabirds, which nest in cliff-top burrows close to where they will find most food. Puffins come ashore by day and

Manx Shearwaters only arrive after dark. Living in a burrow affords both incubating adults and young chicks some measure of protection from predators.

Manx Shearwater research shows that a large colony, in which the young hatch more or less at the same time, is likely to swamp predators with the numbers of young – in other words there will be casualties, but overall a greater proportion survive than if the appearance of chicks was more staggered.

For Puffins there may also be an additional benefit. Around the cliffs and colonies, you can see swirling masses or 'wheels' of flying Puffins. For years this behaviour was not understood, but it now seems likely that this mass of moving birds from the colony is a defence against predators – confusing the enemy and giving the aerial birds a chance to look at the colony from the air without being predated by a large gull or skua.

FRAGILE COLONIES

Perhaps the most famous colonial species was the Passenger Pigeon. It was known for forming flocks of millions, which darkened the skies and gave protection from predators. However, once hunters started to reduce its numbers, breeding success plummeted and extinction was inevitable.

Puffins wheeling in front of their colony – a behaviour that appears to be adapted to confuse predators.

Flocking

It is ironic that our growing public appreciation of Starlings is occurring at the same time that the species is decreasing in parts of its range. It's not so much the individual birds that have captured the people's imagination, but the breathtaking formation flying that they perform as they gather at their evening roosts. Excellent films made for television have brought the drama and the beauty of Starling gatherings – 'murmurations' – into our sitting rooms, without the need to go outside on a chilly autumn evening. But now there are places set up in Britain where you can witness this spectacle for yourself – suddenly people are discussing Starlings, and these birds are climbing the charts of popularity!

WHAT IS A FLOCK?

It might seem obvious that 'flock' is the name given to a group of birds. However, there is more to it than that, because the term is usually, and correctly, used to describe a group of birds that are socially engaged in some way.

One of the wonders of nature – Starlings in their pre-roost assemblies. These assemblies consist of many individual flocks of birds that have been feeding together during the day.

Flocks of migrating Knot are driven off their feeding grounds by the highest tides and flock to their roost to wait for the falling tide and their muddy feeding grounds to be revealed. In this flock there are birds in both breeding and non-breeding plumage.

There is something magical about birds moving together en masse. They seem transformed into one creature with one intent, performing sudden twists and turns in perfect unison – a natural kaleidoscope – often against the backdrop of a setting sun.

Flocks may be large or small. Birds such as Knots form flocks that regularly number thousands, but a group of half a dozen Turnstones flying along the shoreline could also be referred to as a flock. Some birds breed and feed communally and are therefore in flocks for most of the year, but generally in Britain and Europe flocking takes place outside the breeding season.

Often a flock is composed of a single species. Snow Buntings form flocks in winter and feed together along the strandlines in many coastal areas, especially around the North Sea. However, mixed-species flocks are also common outside the breeding season when, for example, families of tits move along hedges and through woodland. These roving flocks attract other species, such as warblers and Goldcrests, which travel with them.

Social styles

Not all species have a social life. Some, such as Great Spotted Woodpeckers, never normally form or join flocks. At the other extreme are birds like Rooks, which seem to always be in flocks and nest close together in noisy rookeries. Then there are some

Social clubs

Groups of six or more Long-tailed Tits form outside the breeding season. Some of these flocks are made up of adults, which may be failed breeders. Other flocks are family groups – parents and offspring – and sometimes two or more families join together.

species, like Pied Wagtail and Lapwing, that are territorial when breeding but form flocks at other time s.

Woodland tit flocks will usually break up in the evenings as different species choose different roost sites. However, in winter on the estuaries, flocks of Knots will feed and roost together.

Individual feeding distances of the birds in the flocks also differs, depending on the species and the habitat. In woodlands the foraging flock may be spread out with several metres between individuals, while in open country, a field or on mudflats, individuals in flocks often feed very close together.

Who's in the flock?

You might suppose that a flock is made up of an equal cross-section of the local population, and often that is the case, with Rooks as an example. However, the demographic of a flock may be more narrow. Some waders migrate in waves, with females first, then males and finally juveniles – so flocks may comprise a single sex or age group. Young Starlings gather in flocks after leaving their nests. Female Chaffinches travel further than males, so some of the flocks are mainly of a single sex.

Finch flocks · Finches frequently join together in mixed-species flocks. The birds roost alone, but flocks form on their feeding grounds each morning. Actively foraging flocks are most obvious early and late in the day, with a rest period around mid-day. Chaffinches, Bramblings, Linnets and Greenfinches often mix together, and will join (or attract) other seed-eaters such as Reed Buntings and Yellowhammers.

While feeding, these different species will mingle together, but in flight they will usually reform into flocks with their own species – except for Bramblings and Chaffinches, which usually remain integrated.

The benefit for the birds in this arrangement is that their preferred food of weed seeds is usually encountered in 'clumps' which may be widely separated. Within a 'clump' there is plenty of food in a small area, which reduces competition, and the many eyes of the flock increase the chances of finding new feeding areas.

There are seasonal differences as well: in summer birds gather at feeding areas and return to their territories, while in winter they move around as a single, synchronised flock with much more social cohesion; they may even roost together.

A scrum of Wrens · Wrens do not normally form flocks, and defend a territory from other Wrens all year round. However, they do allow their territories to break down in cold weather, and at these times prefer to roost communally for additional body warmth. The numbers involved in these mass roosts can be surprisingly large (see page 190).

WHY FLOCK?

Birds would not feed in flocks if it was not advantageous to do so. For a start, more eyes means more chances of finding food. Birds are attracted to where there is plenty of food, and other feeding birds are indicators that there is food to be found.

Younger, less experienced birds particularly benefit from joining a flock. They benefit from other birds' knowledge on where to find the best feeding areas and other resources, and it

Greenfinches nest in loose colonies but feed together in flocks, even in summer.

The numbers game

Flocking helps birds feed more efficiently and reduces the chance of predation – more eyes mean more chance that a threat will be spotted, and each individual bird's chance of being the target of a predator is reduced. These advantages apply both to same-species flocks and flocks with two or more species.

gives them the opportunity to meet potential mates before the start of the breeding season.

Out at sea, feeding Gannets converge on shoals of fish. It is thought that the gleaming white plumage of adult Gannets diving headlong into the water as they feed will attract the attention of other Gannets. Greater numbers of diving Gannets in one spot will help to break up a shoal and may make the disorientated fish easier to catch.

Research on Starlings has shown that their 'pecking rate' increases if a number of birds are feeding together. The reason for this is that each bird can spend less time looking around for approaching danger. However, if there are too many birds for the feeding area, and their individual feeding distance becomes too small, then squabbles break out and pecking rate goes down.

Starlings feeding as a flock on a short grassy field will work the area methodically in their search for invertebrates, but those at the back are at a disadvantage as they follow their flock-mates over ground that has already been searched.

Watching a feeding flock, you will notice a constant 'leap-frogging' going on, as those at the back fly to the front.

Sometimes flocks are more successful at gaining access to food than birds acting alone. In winter, a Mistle Thrush often defends a particular berry-laden tree, as a sort of personal larder. Another passing thrush, such as a Blackbird, may be attracted to this tree but will be driven off by the Mistle Thrush. However, a nomadic flock of thrushes such as Fieldfares is more than a match for a lone Mistle Thrush and may fly in and steal the whole crop!

Danger in numbers?

It is often stated that a bird is safer in a flock, but a flock can be a mixed blessing. Flocks attract predators. Sparrowhawks will attend the evening gatherings of Starlings and pick off any weak birds. Merlins and Peregrines will also visit wader roosts for the same reason.

It is true, however, that an individual bird is safer when there are more eyes on the lookout for danger. There is an additional

A diving Gannet is eye-catching over a long distance, and is thought to provide a visual signal to other Gannets that food has been sighted.

Birds of prey, such as a
Sparrowhawk, can cause panic
among flocks of birds – even
those too large to be their prey.

benefit in the blizzard of wings that greet an attacking
predator and may cause confusion. Overall, a healthy bird is
probably safer in a flock – especially at the centre of a flock.

Birds in flocks can also sometimes turn the tables on a
predator and drive it away. I once watched an Osprey over
Leighton Moss with a 'smoke trail' of Starlings chasing it!

A little help from friends

Flocks of large birds, such as geese, swans and gulls, travel
in 'V-shaped' flocks. Some adopt other formations that are
variations on the 'V': they may be in lines or chevrons. The
birds behind the leader use its slipstream and benefit from
reduced drag.

It is tough on the leaders, of course, but the order in these
flocks regularly changes. With geese, flocks normally comprise
many different families travelling together, which allows for

different adults to take over the leadership from time to time.

Waders and some songbirds fly in dense flocks and achieve much the same advantage without 'V' formations.

Travelling parties

There are other advantages in migrating in flocks. Navigation of the group may help inexperienced individuals find their way to the winter quarters. Also, there are more eyes on the lookout for food at stop-overs.

Many birds, especially wildfowl and waders, leave their breeding areas and go to assembly points, where they form flocks ready for their onward migration – indeed young Swallows and martins gathering on overhead wires in late summer and autumn may be doing exactly this in preparation for their migration.

Flocks of larger birds migrate using thermals – currents of warm rising air – to gain height before gliding long distances. It is uncertain whether there is a social connection between the birds – they may be simply using the same route at the

Large birds come together in vast numbers on well-used migration routes, but it is uncertain whether the birds in these flocks are socially connected.

From the top of one thermal, White Pelicans reunite into small 'Vs' which spread out, watching for one of the groups to pick up a new thermal that can be used by the others.

same time. What does seem certain is that they are attracted to new thermals by the presence of the birds ahead of them, and so there are visual connections, whether deliberate or not.

SYNCHRONISED FLYING – HOW DO THEY DO IT?

The simplest questions are often the most difficult to answer. Observers of murmurations of Starlings or swirling masses of Knots over an estuary frequently wonder which bird is leading the flock and how the unified movements are co-ordinated.

The answer is that there is no leader. Each bird avoids crowding its neighbours, so it maintains the same distance from the birds flying next to it on all sides. Its reactions are virtually instantaneous. As a neighbour – perhaps an older and more experienced bird – changes direction, so too does our bird and so too does the one on the other side and so the message passes swiftly through the flock.

TRADITIONAL FLOCKING AREAS

Migrating birds tend to have favoured spots along their routes where they stop off, often in flocks. Dotterels migrate across Europe to breed in mountain areas. On migration they may stop over in Holland (where, for a time, some started to nest) and in parts of eastern Britain. So regular were their gatherings that sporting events were planned around their arrival. 'Dotterel Day' was in the second week of May, and even today these birds continue to be seen in some of their traditional fields at about this time.

COLLECTIVE NAMES

Over the years local names have been ascribed to flocks of birds. I have already referred to murmurations of Starlings. Most collective names appear to be very old and many come from The Book of St Albans of around 1470–1480, apparently written for or by Juliana Barnes, Prioress of Sopwell Priory, near St Albans. Some of these names are used in heraldry and many have even older origins.

Collective nouns include a wisp of snipes, a murder of crows, a spring of teal, a charm of goldfinches and a gaggle of geese – although a line of geese in the sky is usually called a skein.

Hunting has helped perpetuate some of the old names of groups of birds. A flock of Grey Partridges is called a 'covey'. It may include adults and juveniles.

Songs and calls

As the countryside returns to life after the rigours of winter, woods and fields fill with the sounds of singing birds. Beginning with wistful Robins and strident Wrens, others join in, until it reaches the spring climax that we call the dawn chorus. This chorus of birdsong on an early May morning is one of nature's gifts to us, and it's truly uplifting to go out and savour it. All year round, bird songs and calls help to relax us, they alert us to the changing seasons and remind us of childhood or holidays or other pleasurable experiences. These sounds are ingrained in our culture, our stories and poems, and provide the atmosphere to our films and radio plays. Hearing these natural sounds help us renew our links with the countryside and with wild places.

Song is important for securing a territory and attracting a mate.

Like other birds that live near fast-flowing, noisy rivers, Dippers have a song and calls that are loud and high pitched, within a narrow frequency belt, above the frequency of the water's noise, so that long-distance communication is possible.

WHAT IS SONG?

Physically, vocal song is the sound made by a bird as air passes through an organ in the throat called the syrinx. It works like wind being pushed through two tubes of variable thickness and containing vibrating membranes. Different species make different noises, and no two species – probably no two individuals – sound exactly the same: they form a natural orchestra where every instrument has its own character.

Songs are adapted to be heard regardless of environmental conditions. The shrill piping of a Kingfisher or the high-pitched song of a Grey Wagtail can be clearly heard in their watery homes while lower sounds are muffled by the sound of rushing water. The resonance of a Tawny Owl's hoot in the stillness of the night travels through the wood and far into the countryside beyond.

Not all these sounds have the same purpose. What we generally think of as song, including the hoot of the Tawny Owl, is performed to advertise ownership of a territory. Other, less complex calls may have other meanings. It's not quite a language, but more like a code that we can also interpret if we can discover the key.

The 'Elvis' effect

The better the songster, the more attractive he is to the females. Experiments to slow down birdsong in some species (Sedge Warblers in Europe and mockingbirds in North America) have shown that those individuals with more complex songs are more successful at attracting mates than others with simpler songs.

THE PURPOSE OF SONG

By singing, a male bird is usually staking claim to his territory, and warning rivals to retreat beyond its boundaries, as this is the place that it will use for breeding and also, generally, for feeding. The song also enables the male to attract and retain a mate. There is evidence that some females select males with the best song.

Some individuals have more complex and varied songs than others of the same species. Female Great Tits prefer males with the most song types within their repertoire, and it has also been shown that these males tend to survive better and be more successful breeders.

In some species a male may recognise individual neighbours and may know their preferred song posts, and can tell when newcomers arrive and territorial boundaries probably change.

There may be an additional benefit from song: it signals a bird's fitness. There is growing evidence that predators are more likely to leave strongly singing birds alone as, being fit and healthy, they are going to be harder to catch!

ALL THE WORLD'S A STAGE

What makes a good place from which to sing? In general, singing from high up is good, enabling a song to carry further. The Dunnock spends most of its foraging time on the ground, but when it comes to singing, males will often head for the top of a small tree. Birds with low-pitched, powerful songs are more likely to sing from sheltered places, while birds that give some form of visual display when singing will opt for a more visible spot.

Song posts

Many species select prominent song posts from which they can dominate their territory: a Blackbird on a chimney pot is both highly visible and able to project its song over the surrounding territory. However, others will sing from less obvious perches, and a singing Nightingale in dense cover can be very difficult to observe, even from a few metres. Some song posts are low down; roadside cow parsley for a Corn Bunting, while others are high up, such as the Mistle Thrush singing from the topmost branch of the tallest cedar tree.

A study of American warblers shows that those singing from the highest song posts have higher-pitched songs. Lower-pitched songs and calls travel better through dense vegetation, and there tends to be more dense vegetation at lower levels.

Aerial song

For many species living in open countryside there are no song posts for them to use. The Skylark rises as it sings, getting higher and higher until it is almost lost to human sight. At this high level it continues singing. Later it will descend, still singing, and stop its wonderful cascade of notes only a few metres from the ground. This aerial display is a clear advertisement of an occupied territory, and females seem to prefer males with the longest songs.

Wading birds, such as Whimbrel and Curlew, give bubbling calls as they perform attention-seeking display flights over their moorland homes, adding their evocative sound to a bleak

Species like the Whitethroat use an eye-catching songflight, often highly stylised and unlike normal flight, to boost the impact of their song.

Birds of open country have aerial songflights and calls. The Snipe makes a bleating sound with its outer tail feathers as it zigzags over its territory.

moorland landscape in early spring.

A few species, such as Whitethroat and Sedge Warbler, have both a song post and a display flight. They will sing from a prominent perch, such as the top of a bush, and then occasionally fly up and sing again during a short and rather jerky flight – a good way of getting noticed by females and helping to keep rivals away.

VERSE AND CHORUS

Songs are immensely variable. Many have several distinct phrases. In the case of the Song Thrush, each one is repeated several times, and then another phrase is repeated instead. This happens several times before the original phrase returns. The Chiffchaff and Cuckoo share the distinction of having English names that are imitations of their relatively simple songs.

Many ornithologists have attempted to choose our best

songster, but this is heavily influenced by personal taste. However, the Blackcap is usually high on the list for its pleasant flute-like qualities and clear range of notes, and the rich and powerful song of the Nightingale is legendary.

Not every bird sings sweetly – some produce quite bizarre territorial 'songs'. The Capercaillie, the giant grouse of northern forests, gives a string of grunts and rattles, ending with a 'pop' rather like a wine cork being released; and the Bittern, that well-camouflaged heron of dense reedbeds, gives a deep, breathy 'boom', like someone blowing over the top of a half-filled bottle!

Some species have several different songs. Great Tits can come out with several quite different song phrases. This has confused birdwatchers and led to the axiom that if you can't identify an unfamiliar song in spring, it is probably a Great Tit!

Research also showed an individual Dunnock to have several slightly different songs. The variety of songs made by one bird has led to the development of the 'Beau Geste theory', after the literary figure who tricked an enemy into thinking there were more defenders than there really were – thus rival Dunnocks may be duped into moving to 'less crowded' areas!

Regional dialects

To our ears most songs from birds of the same species will sound similar, but again there is more variety than might be

Boom or bust

The song of the Bittern sounds like a simple 'boom' or 'humph'. These are rare birds and also very secretive. Estimating their population is, therefore, difficult. Researchers from the RSPB discovered that by recording their songs and slowing them down, the pattern made by the recording was sufficiently distinct to be able to recognise individuals and thereby estimate the number of males present – at least those that boomed!

imagined. Some have developed regional dialects, especially when geographical features, such as mountain ranges, separate their populations. Recent research in Switzerland has shown remarkable differences between Skylarks – even those nesting quite close together may have noticeably different songs and this difference increases with distance.

In parts of Europe songs have been found to be different in areas of high traffic noise, which may be a reaction that allows the bird to compete successfully with the additional noise. For example, in one noisy city area in the Netherlands, Great Tits were found to sing at a higher frequency than is normal elsewhere.

Duets

Some species sing duets. Usually the duo is a male and female, but in the case of Long-tailed Manakins of Central America it is two males competing for females. In Europe, Tawny Owls sometimes engage in a type of duetting, with the female promptly responding 'ke-wick' to the hooting song of the male.

A SEASON FOR SONG?

There are many influences on where, when and for how long birds sing. Some, like the Wren and Robin, defend territories for most of the year and sing in most months. There may be a

Up with the lark

Skylarks will sometimes sing a quiet and monotonous song on the ground. In the air they rise to around 50 metres when singing, but some will occasionally climb to 200 metres or more. Most songflights last no more than 10 minutes, although half an hour or more has been observed. Both male and female Skylarks sing, but most output is from the male. Skylarks have complex songs, with between 180 and 460 different syllables. Some of these songs also mimic other local species such as Redshank and Curlew.

short pause during moulting or when there are young to be fed, but generally they are seldom quiet for long – even in the depths of winter.

Most song is associated with the breeding season. Nightingales, which have only one brood, stop singing very early in June and concentrate on rearing their young and feeding up ready for their return flight to Africa. Species such as Blackbird and Song Thrush attempt to rear several broods in a year, and are more likely to continue singing for longer.

It is noticeable that mild autumn weather can prompt a flurry of birdsong. Day length and temperature may be similar to spring, but the autumn song tends to be more subdued without the imperative of breeding. In fact, the Robin's song becomes noticeably slower and more 'wistful', and only after the shortest day has passed does it become brighter, clearer and full of energy again.

Nightingales can frequently be heard singing during the day as well as at night.

NIGHT MUSIC

Most species sing during the day, but a few with nocturnal habits – like the owls – need to sing after dark. The Nightjar is particularly active at dawn and dusk as it feeds on moths and other insects. It is at these times, and especially dusk, that its strange purring, insect-like song can be heard in the still evening air of our heaths and forest clearings. The song rises and falls as the bird turns its head while singing continuously.

A few other species will sing at night. The most well known is the Nightingale, and while it is not exclusively a nocturnal songster, its song is certainly heard at its best after dark, with no competitors to confuse or drown out the song. The song of the male, which arrives back first from its migration, presumably helps the newly arriving females locate potential mates. The clear, pulsating notes are certainly enchanting enough to deserve their romantic association.

Some others will also sing after dark. Sedge and Grasshopper Warblers both do so on occasions, especially while trying to attract a mate. Robins will sometimes break into song on dark nights, even in winter. However, this song is usually triggered by artificial lights and it is this habit that accounts for many mistaken reports of Nightingales singing either in towns or in winter, or sometimes both!

SONG WHATEVER THE WEATHER

Weather can play an important part in song output. For most, wind and rain has a quietening effect, but the Mistle Thrush really lives up to its old country name of 'Stormcock'. It pumps out its loud fluty song from the highest tree tops in the lull just before or after a storm, and sometimes when the wind is blowing strongly.

MECHANICAL SOUNDS

A few species produce a sound without using their syrinx, that is to say without using their 'voice'. These have other vocal calls, but for their 'song' they have evolved ways to make sounds in other ways. Woodpeckers drum their bills rapidly

Giving a hoot

Everyone knows the quavering hoot that is the song of the Tawny Owl. Although it may be heard at many times of year, there are two main song periods – one in late winter as it sets up its breeding territories and again in autumn (October) when it drives out the young of the year from its winter territory.

on a suitably resonant branch to create a distinctive and far-carrying sound. Occasionally another object, such as a pylon, may be used. Generally this is heard in late winter and may be made by either sex. The sound is communication and not connected directly with nest construction – certainly not excavating a nest hole, for which a quite different pecking style is required. Woodpeckers are adapted for drumming with thicker skulls and strong neck muscles.

Noisy feathers

Another remarkable sound is made by the Snipe. This wading bird has unique outer tail feathers which, when spread during the downward curve of its display flight, will cause a vibration rather like the fluttering of a reed in a musical instrument. The peculiar sound produced is known as 'drumming' or 'bleating', and may heard over considerable distances, especially in the evenings and on days with low cloud cover.

Other noises are made with feathers. Male Pheasants will noisily shake their wing feathers and rustle their quills after they crow. Woodpigeons and Nightjars sometimes make loud claps by striking their wings together over their backs as they fly. Usually this noise is associated with courtship, but also, in the case of Woodpigeons, when birds are alarmed. Even the mainly silent Short-eared Owl wing-claps as part of its display.

Some flight feathers are more noisy than others. The humming sound of Mute Swans in flight is especially distinctive, but some smaller birds can also be noisy in flight; for example the Goldeneye's wings make a far-carrying whistling sound. In the case of the Pintail, male and female fly close together in their courtship flight and as a result of their stiff primary feathers touching, a loud rattle can be heard.

Bill-snapping

Some non-vocal sounds may be made with the bill. White Storks returning to their nests clatter their bills in a greeting ceremony. Owls, such as Barn and Snowy, frequently 'bill-snap' at their nests.

White Storks clatter their bill on returning to their nest as part of their courtship ritual.

MIMICRY

Some species have the ability to copy the songs and calls of others and include them in their own songs. Starlings have their own distinctive rather disjointed song of whistles, rattles, clicks and squeaks, but they also add sounds they hear in the local environment. These may be local bird songs and calls, but might also be domestic sounds such as clucking hens or a barking dog, or man-made sounds such as a car alarm. In the 1970s when 'trim phones' were fashionable, Starlings became notorious for fooling people into rushing indoors to answer non-existent calls!

Song Thrushes and Skylarks both incorporate songs of other birds within their own song, but the master mimic in Europe is the Marsh Warbler, which has been shown to copy the songs or calls of up to 84 different species. This may include many that it heard on its migration to Africa, which helps ornithologists work out where it spent the winter.

Mimicry fascinates humans, especially when a bird learns to imitate the human voice. Obviously parrots and mynas are the most expert at this, but historically Starlings were commonly kept as cage birds, and at least some of these learned human speech.

FEMALE SONG

Generally it is the males that sing, but in some species, such as the Robin, both sexes sing. The usual reason for female song appears to be to deter rival females. In the case of the Robin, both males and females hold separate territories in winter, and song is an aid to defending these.

SUB-SONG

There is a form of song that is quite common but frequently passes unnoticed. It has been described as 'sub-song' and its purpose is far from clear. It is quieter and of lower pitch and often longer and more varied than 'normal' song. It appears not to be for communication but more of a 'practice' song, perhaps by young birds before their first breeding season.

COMMUNICATION CALLS

Songs have a very particular meaning linked to breeding, but there are many other noises that birds make that are generally referred to as 'calls'. Familiar ones may be the 'chinking' of a Blackbird alarmed at the presence of a cat, or the persistent squeaks of young Blue Tits urging their parents to feed them.

Most calls have a very precise function and are intended to communicate with others of the same species, but sometimes, as we shall see, even birds of different species can benefit.

Sounding the alarm

A sound that warns of approaching danger is obviously very useful, and most species have alarm calls in their repertoire. A female Mallard with young will quack loudly and urgently when she spots a potential predator. She also splashes off across the water to draw attention away from her ducklings,

This mountain bird, related to the Dunnock, is the only species so far studied in which female song has been shown to attract males. This is important, as male Alpine Accentors help feed the young, and the more the female mates with, the more help she gets at the nest.

An alarmed mother Mallard's calls warn her offspring of trouble, and may also deter a less committed predator from attempting an attack.

which can dive to escape danger. Later there will be a lower volume parental quacking, as she attempts to gather her family together again.

Small birds often react with alarm to a bird of prey overhead, and many species have distinct warning calls in response to this danger. There is, however, a remarkable similarity across different species in the 'aerial danger' calls. A single thin, drawn-out, high-pitched note warns of danger without revealing the position of the bird delivering the call, and its meaning is clear to many different species.

Other alarm calls are more associated with ground predators, especially when 'mobbing'. This activity aims to see off an intruder, or at least draw attention to a lurking presence such as a perched owl. Experiments have shown that the repeated 'chink' alarm call of a Chaffinch in woodland can attract not only other Chaffinches, but 30 other species including Robins, Blackbirds and Great and Blue Tits.

Keeping up with the flock

Another common 'call' helps birds keep together. Long-tailed Tits forage in woodland edges and along hedgerows, and often they go about in groups of a dozen birds or more. All the time, each tiny bird calls continuously, helping to keep the flock together as it moves through the thick bushes. While the

Crow for victory

Pheasants crow and rustle their wings as a territorial signal. They also make these sounds in response to loud noises such as a gunshot or thunder. On 24 January 1915 Pheasants were heard crowing frequently in places as far apart as Norfolk, Lincolnshire, Cheshire, Lancashire and Cumberland. It was only later that it became evident that these birds had been reacting to the air vibrations of the heavy guns of the sea battle known as the Battle of Dogger Bank in the North Sea.

birds are busy feeding it would be all too easy to lose contact without the incessant high-pitched calls. Other woodland species such as Blue Tits, Nuthatches and Treecreepers may also be attracted by the calls and join the flock, benefitting from safety in numbers.

On clear nights in autumn in western Europe, and especially Britain, contact calls can be heard as flocks of birds pass overhead – invisible to the human eye in the darkness. Many of these will be Redwings from Iceland and Scandinavia that have left their northern breeding areas and are migrating for the winter, their contact calls keeping the flocks together in the darkness.

In winter Yellowhammers travel in flocks, both with their own species and mixed with others. Their simple, sharp 'zit' or 'cillip' call help to keep these flocks together.

Feed me

A familiar spring sound is the agitated call of small birds wanting to be fed. These repetitive and penetrating calls help drive parents to greater and greater efforts, but they may have other surprising effects. I once watched a male Wren feeding a family of noisy young Great Tits. It appeared his own mate

was still incubating eggs, and his parental instinct drove him to help this alternative brood – at least until his duties were required at his own nest.

Young Cuckoos benefit from this instinct to feed a noisy chick. The Cuckoo's noisy begging call can be heard at a considerable distance. The call not only prompts the foster parents to continue feeding it but may stimulate other passing birds to deliver food they'd gathered for their own chicks into the Cuckoo's hungry mouth instead.

Complex communication

Each species has its own range of calls for different circumstances. Some have a limited repertoire, while others have a great variety. In addition to the alarm, contact and feeding calls, there are calls associated with roosting, distress, feeding and locating young in a colony.

By promptly evicting the foster parent's own eggs or chicks, a young Cuckoo becomes an 'only child', with a loud begging call that adult birds can't resist.

It is tricky finding your mate or offspring in a crowd – especially when they move around. Many birds can pick out their nearest and dearest by the sound of their voice.

Tell them about the honey

In Africa one species has used a very special call to its own advantage. The Greater Honeyguide's distinctive soliciting call leads other larger animals, from Honey Badgers to humans, to the nests of wild honey bees. The larger animal breaks open the nest and both benefit from the honey inside.

Seeing in the dark

It is now well known that bats 'see' by bouncing their ultrasonic waves off objects, and build up a sound picture that not only allows them to fly in darkness, but also to catch their insect prey.

Birds do not emit ultrasonic sounds, but a few species use echolocation. Cave Swiftlets of South America and Trinidad and Oilbirds of South-east Asia both nest in dark caves, and use clicks and rattles to find their way in total darkness.

THE STUDY OF BIRDSONG

The question of how birds learn to sing has been the subject of study and debate for many years. There are still areas of uncertainty, but great strides were made after the Second World War when the acoustic technologies used for monitoring enemy activity in the oceans could also be used to pictorially display the structure of a bird's song. Sonograms [see illustration] allow scientists to see the 'shape' of the song in terms of its waveform, and how it differs, not only from species to species, but from individual to individual.

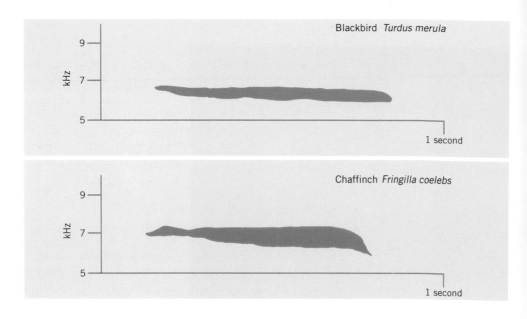

Sonograms are pictorial representations of songs and calls. Some are complex and all differ from species to species, rather like audio 'fingerprints'. These similar-looking alarms by two different species show that that there is inter-species recognition of some alarms, thus increasing the chances of survival.

HOW BIRDS LEARN TO SING

This is a complex issue and whole books have been written about it. A young bird is instinctively equipped to sing a particular song, but it learns the precise phrases and tones from its local environment, and often from its own parents. For many this learning is in its first few months of life and the subdued 'practice' song, known as sub-song, helps develop the full song that will be heard in the next breeding season. In some species this learning continues and more phrases are added as the bird grows older, and the more complex songs of mature males are important for attracting a mate and deterring rivals.

SINGING CAGE BIRDS

Birds have been kept in cages from ancient Egyptian times at least, around 4000 BC. Presumably their songs were appreciated as well as their beauty. From the 16th century, Canaries became increasingly popular for their songs and were imported into Britain from the Canary Islands and Madeira.

Samuel Pepys had a caged Blackbird that woke him between 4am and 5am, but there are also earlier records of Bullfinches

Canaries are still found wild on some of the Atlantic islands, especially the Canary Islands and Madeira. They were first brought to Britain as cage birds in the 16th century because of their attractive song. Selective breeding has developed variation in both plumage and song.

being kept in cages. One account in 1544 mentions Bullfinches kept because of their ability to learn and imitate musical instruments, and even Queen Victoria had caged Bullfinches. Other wild songbirds were also popular as cage birds in Victorian Britain: Goldfinches and Linnets were caught in large numbers, including an astounding record of 132,000 Goldfinches being caught in one year (1860) in Sussex.

Songbirds are now covered by legislation in the UK and cage birds must be bred in captivity and close ringed to prove their origin. In parts of Europe regulations are more lax, and it is all too common to see small birds, especially Goldfinches, in cages in countries bordering the Mediterranean.

WRITING DOWN AND RECORDING BIRDSONG

Describing birdsong has always presented a challenge. As early as 1650 an attempt was made to use musical notation to describe birdsong. The French composer Oliver Messiaen brought this approach to his monumental single work for solo piano, *Catalogue d'oiseaux*, in 1956–1958.

Phonetic descriptions have been commonly used in bird identification guides, but this approach is limited. The alarm

The avian muse

Mozart had a pet Starling that he kept for three years. It could sing notes from his *Concerto in G major*, K. 453, and when it died he gave it an elaborate funeral. Recent research by Meredith West has revealed the piece Mozart wrote about that time which has repeatedly been called *A Musical Joke*, K. 522, has all the hallmarks of the Starling's song: a nine-measure trill, fractured phrasing, continuous repetitions and an eccentric ending – just like the song of a Starling. Perhaps it should be renamed *Requiem for a Starling*?

call of the Blackbird is often described as "a low 'tchook, tchook, tchook'", which may be helpful. Some descriptions go further and suggest phrases that match the song's rhythm and are easy to remember: the Yellowhammer's song, for example, is often written as 'A-little-bit-of-bread-and-no-cheese', with special emphasis on the 'cheese'!

It was not until sound recordings became commonplace that birdsong could be properly captured. The first commercial recording was of a Nightingale, made in Germany in 1910. The first recording of a wild bird was also of a Nightingale,

Ludwig Koch (1881–1974) was a pioneer of early wildlife sound recording and brought birdsong to millions of people through his frequent broadcasts on the BBC. His recordings became the foundation of the BBC's Natural History Library of Wildlife Sound.

accompanied by Beatrice Harrison playing her cello, and was made by the BBC in 1927! Since then recording has improved and sets of bird songs are readily available commercially, especially from the British Library of Wildlife Sound.

Birdsong and music

Many composers, such as Vaughan Williams in *The Lark Ascending* (1914), have tried to represent the spirit and rhythms of bird song in their music. A few, such as Messiaen in his *Catalogue D'Oiseaux* (1958), have gone further, using complex time signatures and tonal structures in an attempt to transcribe birdsong more directly. Some have even incorporated taped recordings of birdsong into their work. They range from Finnish composer Rautavaara, who used the calls of Shore Larks and Whooper Swans in his *Cantus Arcticus* (1972), to English recording artist Kate Bush, whose double album *Aerial* (2005) features the singer imitating various samples of birdsong, including Woodpigeon and Blackbird, and even has a Blackbird sonogram on the cover.

Singing in the rain?

Birdsong appears in the folk memory of the British Isles. The calls of both the Green Woodpecker and the Red-throated Diver have been thought to predict rain — with the woodpecker having an old country name of 'rainbird' and the diver being known as the 'rain goose'. There appears, however, to be no special links between either of these species and the weather.

The juvenile years

What could be more charming than a family of swans, with a line of tiny fluffy cygnets being guarded front and rear by two parents, or a stripy grebe chick nestling among the feathers on a parent's back as it is taken for a ride across a lake? However, these youngsters that are so appealing to us are very vulnerable to predation and other hazards.

TRIALS OF LIFE

It's a sad but essential fact of life that very few young birds leaving their nests for the first time will survive to become breeding adults. Statistically, of a brood of eight young Blue Tits leaving a nest only one will survive to the following year.

As parental help is reduced the juveniles' struggle to find food becomes ever more important. Predators, from cats to crows, specialise in hunting inexperienced youngsters and as summer merges into winter juveniles are faced with declining food supplies, lower temperatures and fewer hours of daylight

Young Mute Swans are often brooded on the female's back, which allows her to continue to feed and regain the weight she lost during egg-laying and incubation.

in which to feed. Only the fittest will be alive the following spring.

Newly hatched

Young birds may start their lives naked, blind, helpless and totally dependent on their parents – this is typical of songbirds. At the other extreme, the megapodes of Australia hatch from their 'compost nests' and are totally independent of their parents from day one – indeed, they may never see their own parents!

Between these two extremes are young birds that follow their parents and are able to find their own food, such as ducklings and most wader chicks. Others, such as partridges, also follow their parents, but have to be shown what to eat. Some, such as rails, follow their parents but are also fed by them.

Some young birds, such as gulls, are covered with down when they hatch but stay within the nest for a time. Birds of prey are also downy when they hatch but are weak and almost immobile, needing to stay in their nests for some weeks. However, young falcons hatch with their eyes open and owl chicks hatch with their eyes closed.

Taking a ride · Riding on a parent's back in the first few days after hatching has several benefits for young water birds. It allows them to keep warm while the adults move closer to

Most young birds begin their lives well protected by their parents, but in the case of Mallards and most other ducks the female alone provides this protection

their food supply without having to leave them to the dangers of the riverbank. It also protects them from predators in the water, such as Pike.

Dive! · Threats to water bird chicks usually come from the land, and in dangerous situations their parents often hurry to get the young onto the water where they can escape more easily.

Young Mallard ducklings feed on the water's surface but can dive to escape danger. When a female Mallard 'quacks' an alarm and flaps across the water causing a distraction, her ducklings scatter and dive, giving at least some a chance to survive. The female then has the task of rounding up the brood, and it is not unusual for the odd duckling to get lost.

Mother Eiders have to run the gauntlet of the large gulls, especially Great Black-backed and Glaucous, as they take their newly hatched ducklings to the sea. The gulls are ruthless, and even work in pairs: one distracting the adult while the other

Young Guillemots jump down from their cliff ledges before they can fly, and swim out to sea accompanied by the males who continue to feed the chick.

snatches a duckling. Once in the sea the ducklings have more chance of escape by diving.

Young auks such as Guillemots leave their cliffs and head out to the open sea before fully grown or able to fly. Accompanied by their fathers, they swim out of sight of land and the unwelcome attentions of the gulls and skuas.

Sibling rivalry · Golden Eagles in Scotland usually produce two eggs, both of which hatch, but commonly only the older chick survives because it kills its younger sibling. This habit has been called 'Cainism' after the biblical Cain who murdered his younger brother Abel, and is not unusual in large eagles. The explanation for the behaviour is far from clear, but it appears to be linked with limited food availability, either when the young are in the nest or when the female is laying. Less food at the laying stage may result in a bigger time difference between the eggs being laid and therefore a greater difference in the age of the young – a critical factor in the relationship between the two chicks.

Among waders, where the chicks generally feed themselves, the young are brooded regularly, but as they grow they only require brooding at night. Some young waders, such as Curlew Sandpipers, are tended by females, while with young Dotterels and phalaropes it is the male that cares for the chicks. In other species both parents are present for a time, but generally the female loses interest and leaves first. Males of migratory species will usually move away to moult before the young start their migrations.

Golden Eagles do not breed until they are four or five years old. After fledging they roam widely, but as they grow older they return closer to their natal site. In Scotland they may travel 100–150km from their nest-sites, while in Norway with more suitable habitat to roam, distances of up to 1,000km have been recorded.

Moult

Whether naked or downy on hatching, the first true feathers a young bird grows form its juvenile plumage. This is generally simpler than adult plumage and more 'fluffy' owing to fewer barbules and hooklets.

The juvenile plumage often offers good camouflage, and may be similar in appearance to that of the adult female or, in some species like the Starling, a totally different plumage

A juvenile male Blackbird is brown like a female and slightly spotted. The late summer moult replaces the body plumage with black adult feathers, but the duller-coloured primary feathers are retained until the following year's moult.

from either parent. Juvenile Robins lack the red breast of their parents and instead are entirely speckly brown, which may help with camouflage, but also prevents territorial aggression from adults.

In many species, especially songbirds, the next plumage is grown during a late summer moult in the same year, at about the time that adults are also moulting. This is usually a partial moult, with some or all of the stronger flight feathers retained until the next season.

In a few species of small birds, such as Pied Flycatcher and

Puppy fat

Studies of young Manx Shearwaters indicate that they accumulate enough body fat to take them from the European nest-sites to the coastal waters of Brazil without the need to stop and feed. One bird completed this journey in about 13 days, which is a rate of 740km per day.

Common Rosefinch, it may be several more years before the full bright adult male plumage is at its best. However, most small birds are indistinguishable from their parents at a year old. This is not the case with many of the large seabirds, which go through a succession of moults into progressively more adult-like plumages over several years. Smaller gulls, such as Black-headed, take two years to attain full adult plumage, but the larger Great Black-backed takes four.

Gannets are extreme among British and European species. Newly hatched birds have bare dark skin, and then grow a white down. The first full juvenile plumage is all-dark, and through a succession of annual moults the bird becomes piebald and finally attains fully white adult plumage in its fifth year. It is assumed that the dark plumage prevents aggression from adults.

LEAVING THE NEST

It is a big step for birds that have spent weeks in a nest to leave and make their first flight. Some species receive no parental help once they fledge. Swifts leave their nest holes and may start their migration immediately. There is one record of a British Swift in Madrid only three days after fledging from its nest in Oxford.

Young Gannets are probably too heavy to fly when they first leave their nests, weighing 30% more than their parents! They make a short glide to the sea and then begin their dispersal from their colony by swimming – leaving their parents behind. Their fat gives them the energy for one or two weeks' paddling, by which time they may be off the coast of North Africa. Around this time they should be able to fly and they need to learn the art of fishing by plunge-diving.

The dark plumage of a juvenile Gannet may prevent aggression from adults. It will take five years to develop full adult plumage. Juveniles start brown and gradually grow more white feathers each year.

The female Great Tit changes her behaviour just before the young leave the nest. She stops brooding them at night, and gradually food for the chicks is reduced, with fewer items brought into the nest chamber as the young are encouraged to take food from their parents at the entrance hole. Soon they are tempted out to get food from adults perched on nearby

The weather conditions at the time young Great Tits leave their nest are critical. Cold and wet weather can cause mass mortality at this time.

branches. Once out of the nest they do not return.

Once out and flying (at first not very strongly), young Great Tits leave the area of their nest and roam the locality in small flocks – sometimes mixing with other families, and even following and being fed by other parents.

A young Golden Eagle leaving its nest has a similar problem, although the nest-site could not be more different. Those in trees and on inland cliffs move from branch to branch or ledge to ledge before launching themselves for their first flight. Those nesting on sea cliffs have no chance to experiment – their first flight has to be for real! Once launched, they choose a nearby perch and continue to be fed.

Those eagles eating carrion will feed themselves earlier than those having to make a kill, but at any rate their parents will continue to feed them for about 75–85 days after their first flight and some continue for even longer.

Flying chicks

Some ground-nesting birds can fly from an early age, to escape danger. Flight feathers grow quickly on young Red Grouse and

Grey Partridges. Grouse can lift off the ground at 10 or 11 days old and partridges at 15 days, even though it will be several more weeks before they are fully grown.

Speed training

Young Peregrines have to learn to kill in flight – a very precise skill. After their first flight it is not uncommon to see young birds diving at each other and generally 'playing' together in flight – all practice for hunting. Soon they start to take prey from their parents as they fly, even turning over in flight to take food from the adult's talons.

Some observations indicate even more training may be given by parents: adults have been seen catching prey and releasing it again near their young, or dropping an injured bird for the fledglings to catch.

Family loyalty

Some birds stay with their parents for many months. Goose and swan families, even those breeding in the Arctic and making long migratory flights, stay together as families while migrating. They winter together, make the return journey together and the young birds may remain close to their parents during the next breeding season.

For most species, the end of the breeding season is the time for juveniles to leave their parents and their immediate surroundings. Juveniles of both sedentary and migratory species disperse in various directions. Depending on the species this dispersal may be 1–10km in the case of Red Grouse or 1,000km by an immature Golden Eagle in Norway.

Young Swallows wander in any direction at first, but after a few weeks they consolidate into a general movement towards the southeast and their migration starts with birds flying 20–30km a day.

Forming a club

Some immature birds that cannot breed in their early years get together with other young of the same species. Herds of Mute

Juvenile Peregrine chasing Wigeon. It takes skill and experience to take prey in flight.

Swans in summer is one common example.

Young Gannets aged up to five years gather on the edge of large gannetries, especially at the height of the breeding season. In these 'youth clubs' much display takes place, and the young birds will visit nearby fishing grounds – all useful practice for when they return as full adults and are ready to breed.

Nocturnal visitors

Manx Shearwaters nest on islands off the coast of western Europe. The young leave their nest burrows under the cover of darkness. A journey of many days takes them to South America – off the coast of Uruguay and Argentina. The following year they will return to fishing grounds close to their original breeding colonies.

In the first year very few of the newly returned juveniles venture ashore. In the second year they land for a short time in June and July, but in the third year they arrive earlier and stay longer. By the fifth and sixth years they are inspecting burrows

Young Kittiwakes disperse and cross the Atlantic after they fledge. They will spend their first winter off the shores of North America feeding on the Grand Banks with Kittiwakes from Canada. They usually spend the next summer there as well.

Fulmars will not breed before their seventh year and may even be 10 years old before they nest for the first time. For most of that time they will have been at sea, although some will make a short return visit to their natal breeding colonies each year.

and by the seventh they are ready to breed.

What brings them back early? The richness of the fishing, which is responsible for the colony is one factor, and the interactions with other shearwaters is another. As they grow older they become more familiar with the island, its burrows and other information they will need for breeding.

Dispersal by juveniles

For migratory species it is not unusual for juveniles to travel further than adults. Young Gannets follow the coast of Africa southwards and may reach tropical waters. Most adult Gannets winter further north than this. The juveniles will not usually return to their home waters until their second or third year and even then they may visit a different colony. By the third and fourth year they join gatherings of other young Gannets near their original colony and prepare to prospect for their own nest-site.

The young eagles we discussed earlier may range over vast areas in their first few years – perhaps 12,000–16,000sq km. However, by their fourth or fifth year, when they are capable of breeding, they have usually settled much closer to where they were reared.

The time between leaving the nest and reaching breeding age will be challenging and full of dangers for every young bird. Those that survive this test will be the strongest and fittest of their kind, and well prepared to rear the next generation.

Survival and longevity

So, how old is the Robin in your garden? Usually this is an impossible question to answer, unless the bird happens to have been ringed as a chick. Whether that Robin is the same one that was there last year is equally unknowable, unless the bird has some kind of unique identifying mark. However, knowing something about the species' usual survival rate gives us an insight into the probabilities involved, and sophisticated research is helping us develop our understanding on this difficult subject.

AVERAGE OR MEAN AGE

For many years most reference books used the average age of birds as a measure of their longevity and indeed this is

Each year millions of birds fledge from their nests, but the vast majority will be dead within 12 months, and many within days of leaving their nests. On average only one of these young Great Tits will survive to become an adult.

Ups and downs

Sometimes there are incidents that have sudden and dramatic effects on bird populations and shorten the lives of many individuals. The harsh winter of 1962–1963 is still one of the coldest on record, and it reduced the populations of many species, especially small insectivorous ones such as the Wren.

The Wren population was suddenly reduced by around 80%. Yet within four years its population had largely recovered, and by the time of the first national *Atlas of Breeding Birds in Britain and Ireland*, surveyed 1968–1972, it was one of the most common and widespread species in Britain and Ireland. Such is the power of recovery in fast-breeding species after a natural disaster.

necessary. How long the 'typical bird' in a population will survive, together with the species' annual reproduction rate, gives us a simple insight into the productivity of the species – a necessary tool to understand population dynamics, which in turn is a tool of nature conservation.

Ringing studies help reveal the lifespan of wild birds. Over the last century a great many birds have been ringed. The rings have been improved and now last longer, particularly on seabirds and large birds of prey where some early data was lost due to wear of the lightweight metal rings. Over time, increasing numbers of ringed birds have been recovered and some fascinating data is being gathered, showing that many species are surviving for much longer than most people would have expected.

Of course the average age of birds is still very low, due to the enormous mortality in their early years. As we have seen, on average only one in eight Great Tits survives to become a year old. A study of Goshawks showed 63% mortality in the first year, 33% mortality among the survivors in their second year

A study showed that, on average, 2% or more of young Song Thrushes may die each day in the weeks after they leave their nest. By the time they are ready to breed the mortality will have fallen to about 1% per week. In general only 20% of chicks will survive to breed.

and 19% in the third year.

It is evident that birds that manage to survive their first year were probably the strongest and fittest of all the 'new recruits' to the population, and will have learned many of the necessary life skills. So, for a time, their life expectancy increases. A study of raptors showed annual mortality was similar year on year once the birds had gained maturity.

SURVIVAL FACTORS

In general, larger birds survive longer than smaller ones, and many seabirds survive longer than land birds. The long-lived seabirds take longer to reach maturity and also produce fewer eggs and chicks annually compared with many other groups of birds. However, as a rule, if populations are to remain stable then all that is necessary is for the adults to reproduce themselves successfully once during their lifetime.

The different behaviour of males and females can affect survival rates. Often males are more obvious and aggressive, but females are more vulnerable on their nests and, in some species, the two sexes have different dispersal patterns once they first become independent, or even different migration patterns as adults. Therefore, survival rates of the two sexes are sometimes different, and research has shown, surprisingly, that among many small farmland species the survival of adult males was almost always higher than adult females.

Another curious survival difference has been noticed in a population of Ruffs. Dark-plumaged males are dominant at leks (dislaying sites) and mate with the most females. Their displays attract 'satellite' males with paler plumage that display less and mate less than the dominant males. However, the survival rate for these satellites is better than that of the darker dominant birds.

DISCOVERING OLD BIRDS

Of course, we can't really know birds' true maximum ages. Only a tiny proportion are ever ringed and, of those, only a minute proportion is found again. The chance of the ones found being the 'oldest ever' is highly unlikely. However, as more are recovered each year, the maximum recorded longevity creeps up – giving us a clue of the potential lifespan of these birds.

There are many factors that can bias the results: obviously birds that die at the hand of man do not live out a natural life,

The benefits of age

For some species, like Kittiwake, older birds in good condition return to their colonies earliest and take the best nest-sites. They also lay larger eggs and rear more young. The average lifespan of a Kittiwake is around 20 years, and the oldest so far recorded was 28 years old.

and it is their rings that are more likely to be recovered than birds dying naturally. Hunting may also influence the time of year birds are killed and sometimes the age of the bird is a factor. A study of the mortality of Goshawks showed young birds were more likely to be shot than adults, as the juveniles wander further and undertake longer migrations.

Small bird longevity

Small birds have a high metabolism, they need to feed frequently and are low on the food chain, being on the menu for many predators. They are relatively short-lived, though their potential lifespan is much greater than most similar-sized mammals. However, to compensate they have large clutches of eggs: a Wren, for example, lays five or six eggs and a Goldcrest six to eight. Both may have second broods and so, theoretically, each year they could produce 12 and 16 young respectively. Unless there is a sudden population explosion, logic tells us the majority of these young birds will be dead within 12 months.

Ringing shows us that the average lifespan of a Goldcrest (Europe's smallest bird) is only two years, but the maximum age recorded so far is seven years. Wrens, only slightly larger,

Same old Blackbird

Your garden Blackbird may well be the same one as last year... and the year before... and even the year before that! Although its average lifespan is three years, the maximum on record in Britain is 14 years, and there is a record of a German Blackbird surviving for 21 years.

also have an average lifespan of two years, and the oldest recorded so far is six years old.

Amazing Robins · Our first example in this chapter, the Robin in your garden, also has a typical lifespan of two years. Its annual mortality has been calculated at 72% for juveniles and 62% for adults. Despite that, the oldest currently on record in Britain was eight years and four months.

In Europe, there is a Polish report of a Robin being killed in its 17th year, and even more surprisingly a Czech record of one being found freshly dead aged 19 years. This demonstrates the survival potential of even very small birds.

Many Wigeon are long-distance migrants and travel tens of thousands of miles in their lifetime. The typical lifespan is three years, but one wild Wigeon is known to have survived for over 34 years.

Big bird, long life

A study of Golden Eagles in Scotland concluded that an average lifespan would be 20–30 years, but with individuals perhaps capable of surviving to around 39 years. This estimate is given credibility by two European recoveries; one of a 25-year-old eagle in France and another of at least 32 years in Sweden. There was also a record of a 42-year-old bird in captivity – which while not a completely fair comparison, helps confirm the overall picture.

Mute Swans are well known for their faithfulness to a partner and to their nest-site, and those in the south and west of their range seldom travel far. Their normal survival rate is about 10 years, but in Britain the oldest ringed bird found so far was 26 years old.

The migratory swans – Whooper and Bewick's – live for almost as long, with an average of nine years. Despite their long regular migrations, Whooper to Iceland and Bewick's to Siberia, their maximum recorded ages are impressive, with one Bewick's living for 20 years and a Whooper for 23 years.

Long-lived travellers

Migration holds many dangers, both natural and man-made – especially for those species travelling long distances and crossing open oceans and deserts. The casualties are high but

Herring Gulls are long-lived, with
a typical lifespan of about 12
years, and one known to have
reached the age of 30.

remarkable numbers survive year after year.

Pied Flycatchers make non-stop crossings of the Sahara
Desert to reach their wintering areas in sub-Saharan Africa.
While their average lifespan may only be two years some birds
live a lot longer – including a British bird that lived to be nine
years old. Because of its route to winter quarters it will have
made the desert crossing 18 times.

As we discover more about migration routes, we are coming
to realise just how far some birds travel during their lives.
Whimbrels from Iceland travel to the coast of West Africa and
back each year. Their average lifespan is 11 years and there are
birds that have made the journey in 16 successive years.

Wildfowl are also great travellers, and long-lived birds.
Pochards that winter in western Europe, migrating from
eastern Europe and Central Asia, have been known to live for
22 years. Shelducks have an average lifespan of 10 years and
have been known to survive until 24 years old, and Wigeons

moving south through Europe each year have a normal lifespan of only three years but have been known to reach the ripe old age of 34 years.

Perhaps not surprisingly, geese, many of which are also long-distance migrants, also have some impressive records. Their average lifespan is six to 11 years, but the current record for Greylag is 18 years, Barnacle 26 years, Brent 28 years and Pink-footed a remarkable 38 years.

Record beaters

Seabirds include some of the longest-lived species. The Arctic Tern may live for more than 30 years and each year it makes its 70,000km round trip from its breeding grounds in the Arctic to its wintering area in the Southern Ocean, off the ice shelf of Antarctica.

Manx Shearwaters are charismatic summer visitors to islands in the North Atlantic. Although nocturnal, and wintering at sea, they have been studied for many years and ringed. Some early rings became badly worn so that some early records are now lost, but in recent years there is a record of at least one bird that has reached the age of 55, making it the oldest European bird.

The larger relatives of the Manx Shearwater, the great albatrosses of the Southern Ocean, can survive even longer. A Laysan Albatross which was ringed as an adult in 1956 was still going strong in 2011, making it about 60 years old.

AGE OF FIRST BREEDING

Linked with longevity is the age at which birds start to breed. Most small songbirds, with relatively short lives, are ready to breed the year after they hatch, but for some other species it may be considerably longer. An Oystercatcher will not breed until its fourth year, and produces three or four eggs a year, but it has an expected lifespan of 12 years and a record longevity of 36 years. Fulmars will not breed until seven to 10 years old, and lay only one egg a year, but can be expected to live for 40 years.

Storm Petrels are not much larger than Great Tits. Apart from coming ashore to nest they spend their lives on the open seas. There are records of these tiny birds surviving for more than 30 years.

Population regulation

Birds are such a familiar part of our world that we tend to take them for granted. Ducks on a park pond, gulls on a harbour wall and Blue Tits visiting our bird feeders in winter: they have always been there and always will be... or will they? Bird populations are seldom stable – there are always species increasing and others that are decreasing. Some of the causes will be as a result of man's impact on the environment, but there are also many natural causes for rises and falls.

POPULATION REGULATION

We shall look at two broad categories of constraints on birds and the way they live. The first are those of natural origin – those environmental factors that have been influencing the populations of living things since life first appeared, and which in fact have driven the evolutionary process and generated the

Yellowhammer is one of the seed-eating birds affected by modern agriculture. The loss of stubble has reduced its food in autumn, and lower insect numbers mean less food is available for its chicks in summer.

diversity of life that we see today. The other category comprises the man-made factors, both harmful and beneficial, on all nature and birds in particular.

Many birds are at the top of their food chains. They are not affected by environmental changes any more or less than other creatures, but they are usually more obvious than other groups of organisms, and bird populations can be reliable indicators of the health of the environment.

The amount of food available and the number of safe nesting places limit the expansion of a population. Surprisingly, predators may not have a major impact on the population growth of their prey, as they mainly take the surplus that would not otherwise have found enough food or nest-sites – leaving more for others, which feed and breed more successfully with less competition around.

Birds – especially the smaller and faster-breeding species – have the capacity to recover quickly from a natural crisis. A population that has suddenly been reduced by a 'one-off' environmental episode usually has the ability to recover, such as Wren populations after cold winters. The survivors can benefit greatly from reduced competition, and enjoy better breeding success than they would when the population is at normal levels. Obviously this has not always been the case, and some bird species have not recovered from environmental changes in the past. Extinction has been the inevitable result.

Climate limits

We know that great changes in bird distribution followed the last Ice Age as species moved north, taking advantage of new feeding and breeding sites as they followed the retreating ice. In much the same way, some southerly species such as Little Egret have spread north more recently, presumably in response to climate change.

Weather

Weather continues to play a limiting role in the life of birds, especially away from the tropics where there is greater

The Wheatear is one of the species that is assumed to have originated in Africa and spread north after the last Ice Age, as the ice caps retreated. Today all Wheatears return to Africa for the winter, even those that breed north of the Arctic Circle.

Little Terns breed on beaches close to the sea where their nests can be washed away by high tides. Their breeding success can also be jeopardised by tourists.

variation in weather patterns. Cold winters, wet springs, droughts, floods and gales can all reduce bird populations, but occasionally bad weather creates conditions that help birds. Flash floods can, for example, provide unexpected new feeding and even breeding areas for some species. More detail about the effects of weather on birds will be considered later.

Tides

Birds are generally able to take advantage of the rise and fall of the tides, and many use the intertidal area for feeding, but some species breed near the strandline, which carries a risk. Little Terns frequently nest on or close to the high-tide line of seaweed and other detritus and are in danger from the monthly 'spring' tides, that are higher than other tides, and also from strong onshore winds that drive waves up a beach and swamp nests and eggs. Exceptionally high tides can devastate these colonies.

Disease and natural poisoning

There are a number of viral, fungal and bacterial diseases that affect wild birds. Outbreaks are often linked with food shortages and wet weather but are quite natural within the population. Usually they pass unnoticed, and have little or no appreciable effect on bird populations.

Trichomonosis

The Greenfinch population in many parts of the UK has been noticeably reduced by trichomonosis. Greenfinches feed in flocks, and social feeding helps spread this type of disease. The spread of disease in gardens is a reminder of the importance of sensible hygiene around bird tables, feeders and places where birds drink. Anyone who feeds or provides water for garden birds should clean feeding areas and bird baths regularly.

Toll of the tide

'Red tide' appeared in 1968 near the Farne Islands, off the coast of Northumberland, and 80% of the Shag population died in the space of a few days – the species had, until then, been increasing at a very fast rate.

One infectious disease that has had a noticeable effect on bird numbers is trichomonosis, which emerged in 2005 among garden birds (having previously occurred in pigeons and doves) and also the birds of prey that fed on them. The most vulnerable species is Greenfinch, and the population fell by 30% in some places. Chaffinches were also affected. It appears this disease is transmitted either by the birds feeding each other in the breeding season or through contaminated food or water.

In 2007 there was widespread alarm at the spread of avian influenza (bird flu), itself not uncommon in birds, especially captive birds. The strain H5N1 seemed particularly virulent, spread quickly and led to human deaths in Asia. This resulted in restrictions on the movement of birds in general and a ban on the import of caged wild birds into some countries, including the UK.

Periodically there are algal blooms in either fresh or seawater. These proliferations can be caused by human activity or be entirely natural – even the result of warm weather. At sea, the water often turns red, colloquially known as 'red tide'. The algae produce a toxic nerve poison that affects birds and mammals. In tropical waters another species of algae can de-oxygenate the water and kill large numbers of fish.

Predators

A few species of birds are 'top predators'; they hunt but are seldom hunted. However, their populations, like those of other animals, are often regulated by the available food supply. Snowy Owls breed in areas where the commonest small mammals are lemmings. The lemmings have their own reproductive cycle, and in some years are extremely numerous but in others become scarce. In years when lemmings are numerous the size of the Snowy Owl's clutch is larger and more chicks survive.

There is a reversal of fortune for other Arctic breeders. In years when lemmings are scarce the Arctic Foxes turn to wader eggs and chicks for food, and the productivity of these ground-nesting birds goes down. One study showed a 40% increase in young waders (Turnstone, Knot, Bar-tailed Godwit, Curlew Sandpiper and Little Stint) reaching their winter quarters in South Africa in a year when lemmings were at their peak. Thus the breeding success of the humble lemming can be monitored at, almost, the other end of the earth.

The economics of ecology

There are two species of lemming in Europe: the Norway Lemming and the Siberian Brown Lemming. Lemmings are prey for Arctic predators such as Snowy Owl and Arctic Fox. The lemming breeding production peaks approximately every three years, and this can have a marked effect on the breeding success of many Arctic breeding birds, not just those that feed on lemmings.

It has been estimated that a brood of nine young Snowy Owls will eat 1,500 lemmings between hatching and fledging.

Red Grouse are limited by the amount of moorland available and the quality of the heather. Regular burning stimulates new young shoots, which provide food for the grouse.

A place to feed and breed

The most obvious limiting factor on a population is food availability; if food becomes scarce then fewer eggs will be laid and fewer chicks will survive. The more the population expands the more competition there is for the food available.

There is also competition for territories. The strongest and fittest birds take the bigger and better territories. Red Grouse breed exclusively on moorland, but those that are forced to set up territory on the fringe of moors may not succeed in attracting a mate and breeding. Therefore the amount of habitat can also limit the expansion of a population.

Accidental death

Some birds die by accident. Often these are inexperienced juveniles and many accidents relate to humans in some way: collisions with cars, windows and overhead cables.

It is relatively unusual for a bird to die by a natural accident – some are hit by lightning in a thunderstorm or become tangled in vegetation, and some young birds become separated from their parents and die from lack of attention. More common are deaths by drowning or as an indirect result of predation, such as young birds starving following a parent being caught by a predator.

The Skylark, a species that presumably evolved on the steppes of Central Asia, is now at home on farmland in the UK.

MAN-MADE HAZARDS

The impact of human activity on bird populations is enormous. Some of the changes date back to early human history. For example, when woodland cloaked much of western Europe there would not have been much room for Skylarks, but with the loss of woodland, and the development of open fields for agriculture, the Skylark, once a bird of steppe and plain and semi-desert, became the epitome of the open countryside in many parts of Europe.

Wet river valleys, marshes and bogs were drained for agriculture and gradually many wetland species were driven to the brink of extinction. Cranes disappeared from Britain and Corncrakes, once familiar across the countryside, retreated to the Western Isles of Scotland and to Ireland. The Bittern, Spoonbill and Marsh Harrier are now restricted to tiny fragments of their original wetland habitat.

Hunting

Since the invention of tools, people around the world have hunted wild birds – for food and other reasons. Many species that we consider rare today were once common in a more 'natural' landscape. The birds shot, such as Pheasants, are managed to ensure their populations remain healthy and produce a shootable surplus, but there have historically been local and global extinctions of species due to over-hunting for food.

Large birds of prey have been killed in huge numbers to protect gamebirds and livestock. Birds of prey and other large, dramatic species were also frequently killed by 'sportsmen' and were beloved as skins by trophy hunters. Owls, Ospreys, harriers, eagles and kites all suffered huge losses. All birds of prey in Britain are now protected by law, but Cormorants and fish-eating ducks such as Goosander may legally be killed under certain circumstances to protect fish stocks.

Hunting can also have positive effects, with large areas of the countryside managed for game but offering good habitat for other species too. Moorland is managed by burning areas in rotation, to improve the heather for Red Grouse. Farmland retains hedges and copses for nesting gamebirds such as Pheasant and partridges, but also provides an ideal habitat for many other species such as Blackcaps and Whitethroats.

Second chances

The Capercaillie was hunted to extinction in Scotland in the 1770s but was reintroduced in 1837, initially for sport. At first it thrived, but in recent years the population has been falling and it is now a species of conservation concern in Britain, with only 1,200 individuals.

Exotic alien

Mandarins are beautiful ducks
from Eastern Asia and were
brought to Britain to add
a touch of glamour to park
lakes, but they soon escaped
into the countryside and
established self-sustaining
'wild' populations.

Introductions

Many non-native birds have been added to our countryside
– often for sport. The Pheasant has been long established,
probably by the Normans, as a gamebird, but its numbers are
added to by the million each year for shooting. Red-legged
Partridges have been introduced to parts of Britain and
western Europe for the same reason.

Species such as the House Sparrow benefitted from the
activity of humans and nested in roofs of houses and fed from
the products of agriculture. The species was taken all round
the world on ships and gradually colonised every continent
except Antarctica.

Other species have been introduced as if on a whim, or
completely by accident. Little Owls are a charming European
species introduced to Britain in the 19th century for no
particular reason that I can discover. Ring-necked Parakeet,
a popular cagebird, has established populations in London,
Brussels and other European cities following escapes from
captivity.

Some introductions have had unforeseen consequences. The
Ruddy Duck, a common species from North America, escaped

into British wetlands from wildfowl collections, but it soon adopted its migratory tendencies and started to appear in mainland Europe where, by interbreeding, it threatened the endangered White-headed Duck. A programme to exterminate the Ruddy Duck in Europe is under way as I write.

Conservationists have also reintroduced species that had been lost. Red Kites now fly again in places where they had been exterminated centuries before. White-tailed Eagles glide again along the cliffs of the Hebrides where once they had been lost. However, a reintroduction will only be successful if the factors that caused the species' disappearance in the first place have been removed.

Pesticides

One of the most shameful episodes of the 20th century involved birds. In the battle to protect farm crops from insect pests, special pesticides were developed. DDT was in use from the 1940s and followed by other toxic organochlorines including aldrin and dieldrin. These had major repercussions: they stayed in the environment for many years, and they accumulated in animal bodies to be passed up the food chain,

Not so marvellous medicine

The White-rumped Vulture may once have been the most common large bird of prey in the world, but in the 1980s it started to decline rapidly across India, Pakistan and Nepal, and by 2000 had virtually disappeared from India. The cause was eventually found to be a drug, diclofenac, used as an anti-inflammatory treatment for livestock, especially cattle. Vultures eating dead livestock were poisoned by the drug. The Indian government outlawed the drug in 2006, and a captive breeding programme is under way to reintroduce the vultures back into the wild – but it will be a long hard struggle, and very many years before they return to anything like their original numbers.

The natural range expansion of the Collared Dove is remarkable. It spread from India to Turkey. It then colonised Hungary in 1932, Austria in 1943, Germany in 1945, the Netherlands in 1947, Denmark in 1950, Sweden in 1951, France in 1952, Britain in 1955 and Iceland in 1971!

reaching their highest concentrations in the top predators. Their effect was to interfere with breeding, so shells broke in the nest before hatching as a result of eggshell thinning.

The result was a rapid decline of birds of prey from the 1950s to the time the chemicals were restricted, which began in 1962. The species most affected were raptors such as Sparrowhawk and Peregrine, but many other farmland birds also suffered population declines at this time.

Changing agriculture

Modern agriculture has affected many wild bird populations. In the UK the Corncrake and the Grey Partridge have retreated, Skylarks are less numerous and Turtle Doves have disappeared from many hedgerows.

Most countryside is managed to a greater extent than in the past, with trimmed hedges, neat verges and fewer 'wild corners' where weeds and insects can thrive. Modern harvesters catch all available seeds, which are transferred to hygienically sealed storage. Gone are the days when birds could depend on spillage and open barns for free food.

In parts of Europe, dead livestock was once left out on the hills, where it was eaten by foxes and birds of prey such as vultures and kites. Modern standards now prevent this – resulting in some declining vulture populations now having to be fed artificially.

Some major declines were inevitable when the sowing of cereals switched from spring to autumn. This apparently small change had far-reaching consequences. Traditional stubble fields were replaced with lawn-like fields of young cereals, providing little or no food for small birds in winter, and lush growth in early spring reduced the places where Skylarks and Lapwings could nest and rear their families.

Grants and stewardship schemes have been introduced to help, and some species are starting to recover. The introduction of random small bare areas in cereal fields have provided safe places for Skylarks to feed and nest, but the population in many areas remains low.

Pollution

Pollution has hit many species over the years. The biggest and most publicised incidents have involved oil spills in the sea, either as a result of shipping accidents or from accidents around oil fields. For many years ships discharged oil into the sea, which led to ongoing oil pollution incidents.

Seabirds are badly affected by becoming oiled, and by reduced food supplies due to the pollution. Tighter controls on shipping and better legislation to protect the sea help, but the possibility of an accident is an ever-present threat.

Fresh water can also become polluted. Major spills of chemicals make the news now and again, but run-off of nitrates from agricultural land into waterways has been very common and can seriously upset the ecosystems.

Collateral damage

Accidental killing of birds by human actions is common. Any driver will be aware of road kills. This may affect disproportionate numbers of particular species, such as Barn Owl, which frequently hunts roadside verges for its prey.

Dolphins became a key cause for conservationists when it was discovered how many were drowning in fishing nets. Birds also suffer in this way, with diving species such as Guillemots getting caught while fishing.

Lapwings have disappeared from many fields in England and Wales due to increased field drainage and autumn-sown cereals.

The trouble with oil

Oil pollution of the sea is not new. As early as the 1920s the RSPB was campaigning for cleaner seas around the British Isles, but it took major incidents such as the wrecking of the Torrey Canyon off the Isles of Scilly in 1967 which killed 15,000 birds, the wreck of the Braer off Shetland in 1993 and the grounding of the Sea Empress off Pembrokeshire in 1996 to really bring this issue to public and Government attention.

Hobbies have spread north in Britain in recent years. This may be a result of climate change linked to increased food in the form of dragonflies.

Even more alarming is the number of seabirds, especially albatrosses, caught by long-line fishing vessels. Birds take baited hooks as they are played out from the back of fishing boats and drown as the hooks are drawn under water. In 2000 BirdLife International estimated that 100,000 albatrosses a year were being killed in this way – a totally unsustainable number, threatening 18 of the 22 species with extinction.

Climate change

Whatever the causes of climate change, it is already having an effect on bird populations.

Warmer summer temperatures will draw species northward to new feeding and breeding areas. In recent years western Europe has seen Serin, Cetti's Warbler, Little and Cattle Egrets apparently spreading north. Higher temperatures, however, will have an adverse effect on upland birds such as the Dotterel in its mountainous, tundra-like habitat.

Some species in the British Isles are breeding earlier as a result of milder spring weather and, presumably, more food being available. This may not be a problem, unless by hatching earlier the young miss the main peak of their summer insect food – which could reduce their breeding success.

If fish populations relocate because of rising sea temperatures, the seabirds that depend on them may have to travel further to find food, and this will affect the survival of their young. Rising sea temperatures also contribute to sea-level rise, which affects coastal nesting birds – with more of their feeding and breeding areas being covered by the sea more frequently. Some freshwater coastal marshes will be swamped by sea water, thus changing their ecology and their bird populations.

THE FUTURE

Over-exploitation of our planet's resources is bad news for us as well as wildlife. Now is the time to take our lead from the natural world and learn how to live sustainably. If we can achieve this, the result will be a happier future for ourselves as well as the natural world.

   As we come to understand birds and all nature better, we can try to recreate some of the environments that have been lost. This offers exciting opportunities, with projects already under way in the UK to create new marshes, reedbeds, woods and heaths. These places can provide opportunities for people to experience and enjoy nature, and they are part of a vision for a brighter future with a countryside once again full of birds and wildlife. Only time will tell how successful this will be.

With fossil fuel supplies running out, we need to find sources of alternative energy. The siting of wind farms is critical. In some locations they will damage the wildlife importance of the area and harm wild bird populations. Careful positioning can prevent much of this damage.

Migration

There are many wonders in nature, but one of the truly remarkable sights is of small birds arriving on our shores after completing a journey of hundreds or even thousands of miles. They may have crossed wide stretches of open sea, and in some cases travelled from one end of the earth to the other. These annual migrations involve millions of birds every year, and yet their scale and complexity is largely hidden from us.

INCREDIBLE JOURNEYS

The marvel of migration is not just the distance travelled, but the powers of navigation that take a Swallow from a farm in western Europe to the southern tip of South Africa and back,

Swallows are visible day-flying migrants, which generally stop and roost at night. It takes them about six weeks to reach South Africa from western Europe, and about four weeks for their spring return.

Female Chaffinches generally travel further than males, and so flocks of females are more numerous in the southwest of their winter range, especially in Ireland. Carl Linnaeus noted in the 18th century that a greater proportion of males than females remain in Sweden over winter. He gave the species its scientific name of *Fringilla coelebs* – 'coelebs' being Latin for 'bachelor'.

not only to the same country and the same locality, but back to the very same barn where it nested the previous year. A young Kittiwake crosses the Atlantic to winter off the coast of Canada, and returns two years later to the very same Scottish colony where it hatched as a chick.

As science allows us to further explore this aspect of ecology, we marvel all the more at the navigational skills and the physical endurance of birds as small as a Goldcrest (a mere 9cm from bill to tail-tip) crossing the North Sea, or an Arctic Tern flying from Arctic to Antarctic and back again year after year for up to 30 years.

WHAT IS MIGRATION?

At its simplest, migration is a journey along a broadly predictable route from one specific place to another, to return again at a different season. Migration also applies to species where populations in some areas travel but others do not. Chaffinches from Scandinavia move south in autumn. Large numbers arrive in Holland and Belgium, with many crossing the North Sea and arriving in Britain and Ireland. Here they join our local Chaffinches, which are resident all year.

There are many other local bird movements that are not generally regarded as migrations, as they do not follow any particular route. Many juveniles leave their breeding areas and

Winter wanderings

European Rooks are long
distance migrants. Elsewhere
many leave small local rookeries
and form larger woodland
gatherings. One gathering in
Scotland contained 65,000
birds. They return to their own
rookeries in January or February.

disperse widely. Starlings will abandon their spring territories
and form flocks with other Starlings to forage in fields and
pastures.

Of course, birds are not the only creatures to make amazing
journeys. The great whales travel from tropical to Arctic waters
and back each year, Blue Wildebeest migrate across the plains
of Africa in response to the annual rains, and a succession of
generations of Monarch butterflies fly from Mexico to Canada
and back again to Mexico – roosting in the same trees used
by their parents or grandparents. A female Common Toad
will crawl a mile or more from her hibernation burrow to find
her breeding pond, often carrying a mate on her back! But
migrations made by birds must be the greatest in the world.

WHY MIGRATE?

Each autumn, 186 bird species from Europe and Asia
migrate to Africa. Populations are largest after the breeding
season, and an estimated 5,000 million birds make this
journey each year.

The conditions are gruelling and mortality is enormous, but

a species as a whole benefits from these seasonal movements. Migratory behaviour evolved during and after the last glaciations, as birds spread north to newly available feeding and breeding grounds, only to return each year to familiar, reliable, food rich wintering areas. In short, migration allows birds to move to where food is most abundant and leave when it becomes scarce, and these journeys ensure the survival of the species.

There are various triggers for migration: declining supplies of food such as insects, fewer hours of daylight in which to forage, and lower temperatures leading to snow and ice making food less accessible.

STUDYING MIGRANTS

In ancient Egypt there are tomb paintings of migrant Greylag, White-fronted and Red-breasted Geese, which would have flown there from further north. While the time of their arrival and departure would have been known, their breeding grounds would have been a mystery.

The Biblical book of Job includes the passage "... the hawk grows full fledged, in time to spread her wings for the southward journey?", indicating that at this time the direction of migration was also recognised.

Snow Buntings breed on Arctic tundra, further north than any other songbird, and are forced to migrate when snow and ice covers their food. The larger males tend to stay closer to their breeding sites and at higher altitudes than the smaller females – many of which winter around North Sea coasts.

Manna from heaven

Quails migrate from Europe to Africa and 'falls' (the name used to describe migrants appearing en masse) occur from time to time around the Mediterranean, depending on weather conditions. The Bible has a record of what was probably a fall of migrant Quails, which helped feed Moses and the Children of Israel in the desert "A wind... brought Quails from the sea, and let them fall by the camp. And the people stood up all that day... and they gathered the Quails."

Rings are uniquely numbered, lightweight bands that are fitted around a bird's leg. In Britain this research is controlled by the British Trust for Ornithology (BTO) and can only be carried out by trained and licensed 'ringers'. The recovery of the rings from corpses, or birds that are re-trapped elsewhere, provides much of the data that helps explain bird movements.

In the 7th century St Isidore of Seville wrote, "The swallow catches and eats its food in the air. It crosses the sea and winters there," yet despite these observations there persisted, for many centuries, stories of birds hibernating. We now know a few species may become torpid (inactive and with a significantly reduced metabolic rate) for a short time in bad weather, but extended torpidity resembling hibernation is only known for one member of the nightjar family.

During the last 200 years much has been learnt about bird migration, at first through field observations and then benefitting from the development of 'ringing' (known in North America as 'banding') for scientific research. Bird ringing first started in Denmark in 1899.

Radar during the Second World War recorded 'angels' on the screens. These were first thought to be enemy aircraft, and only later were they identified as birds, or rather flocks of migrating birds. Using this information, David Lack made ground-breaking studies after the war that demonstrated the value of radar in monitoring bird movements and migration.

Even more recent advances in technology have resulted in the development of transmitters that are small enough to be attached to birds without impeding natural flight and other behaviour. These devices allow day-by-day monitoring of bird movements. Geolocation systems (GLS) and global positioning systems (GPS) are now also used, and can locate a bird to within 20 metres.

The most recent development in the field of migration research is the analysis of particular chemical isotopes stored in birds' body tissues and feathers. Because these are derived from what the bird eats and drinks, analysis of them allows researchers to accurately establish the origin of migrants.

DIRECTION

It is well known that Swallows leaving western Europe fly south to Africa, with those from the British Isles travelling furthest, to Cape Provence – almost the tip of South Africa. Less well known is the migration of birds from further north, especially Arctic regions, which are forced south after breeding and visit western Europe either for the winter or to feed here before continuing their migration.

While most migrants tend to move north–south, there are other variations. Pochards from eastern Europe and Russia move westwards in autumn and arrive on lakes in Britain and western Europe. These are birds escaping the cold winter of the continental interior.

TIMING

As soon as conditions allow, migrating birds move into their breeding habitats. The timing is critical: arrive too early and there might be insufficient food, but too late may mean the best territories are already occupied. Generally males arrive ahead of females, and older, more experienced males will claim the best sites.

Migrating birds can cover 50–250km per day. Some species, such as Swallows, pause at favoured spots for a day or more on their journeys. Their migration may take six weeks or longer, depending on the weather conditions they encounter on their routes.

Most seabirds disappear out to sea after breeding. Many move south to warmer tropical or sub-tropical waters, while

Bachelor party

In autumn, flocks of mainly male Pochards gather in western Europe and southeast England. They have come from the Baltic countries and from Russia, with some travelling from as far east as Moscow and beyond. Males migrate ahead of the females, settling first in suitable areas for moulting and then moving on again. In general males tend to winter further north than the females, and return to their breeding grounds earlier.

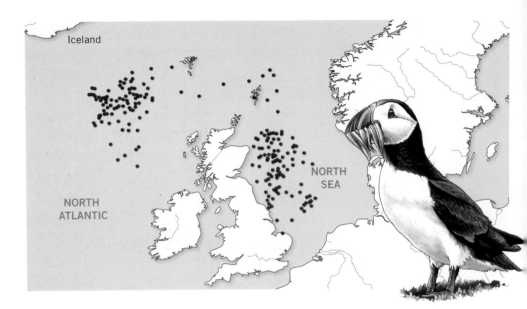

Puffins leave their clifftop breeding colonies as soon as they have finished breeding and will spend the rest of the year out at sea in the North Sea and the North Atlantic until they return to breed the following spring.

others, like the Puffin, may move far out into the North Atlantic. Juvenile Kittiwakes from Europe mostly cross the Atlantic in a westward direction, to spend a year or two in the coastal waters off North America and Greenland.

It is not unusual for young seabirds to remain out at sea until they are ready to nest. Some will make short visits to their natal colonies during the breeding season.

LOOP MIGRATION

It is common for birds to retrace their routes between summer and winter sites. Others, however, use different strategies. Loop migration describes a route that has a different return path to the outward path. Garden Warblers studied in Germany moved to Africa via Spain, Portugal and Gibraltar in autumn, but returned via Scilly and Italy in spring. This annual anti-clockwise journey allows them to take advantage of the prevailing winds at different times of year.

HOW FAR TO FLY

Some migration can be over quite short distances. In Britain,

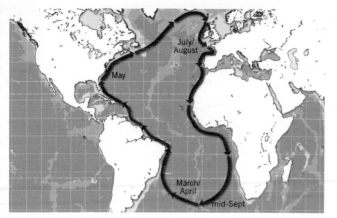

Atlantic tour

Like many seabirds that breed south of the equator, Great Shearwaters, which nest on islands in the South Atlantic, travel north after breeding. Their journey is a great ellipse around the Atlantic. First they travel north along the American coast, then change direction south of Iceland and start to fly south. They pass the British Isles in late summer and continue on, along the coast of Europe and Africa, and back into the South Atlantic in time for the next breeding season.

small insect eating birds from the uplands, such as Meadow Pipit, can simply move down from the hills to winter on lowland or coastal areas where more food is available. This is known as altitudinal migration.

New technology has given us fresh insights into the duration and routes of long-distance migrations. The first to hit the international news was a Bar-tailed Godwit that flew non-stop from New Zealand to North Korea – a journey of 10,400km which took seven days. More recently, Turnstones fitted with geolocators have been recorded flying 7,600km from Australia

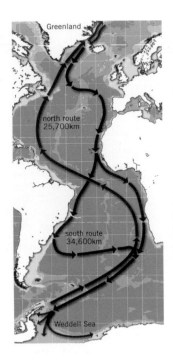

Sun-seekers

One of the most remarkable migrations is that undertaken by the Arctic Tern. In the course of its annual migration, from its Arctic breeding grounds to the Southern Ocean, off the coast of Antarctica, each tern covers 70,000km and sees more daylight than any other creature, experiencing two summers each year! The course it takes is not straight, but is 's' shaped across the Atlantic Ocean, to take advantage of the prevailing winds. This incredible journey may be made many times, as Arctic Terns can live more than 30 years.

to Taiwan in six days without stopping. There they feed before moving on to China, from where they make a further 5,000km flight to Siberian breeding areas. Their return journey is usually by a different route, again with only a few stops.

Small birds can also cover impressive distances non-stop, with Sedge Warblers apparently able to cross the Mediterranean and the Sahara Desert in a single flight – a journey of 1,500–2,500km.

Great sea crossings

To reach their breeding grounds, many birds seem to set themselves impossible targets. All Wheatears spend their non-breeding season in Africa but nest in Europe, Asia and North America, as far north as Greenland, up to beyond the Arctic Circle. This requires long flights, including sea crossings.

In spring, Wheatears island-hop from mainland Europe to Britain and then on to Iceland, then Greenland and some even reach Arctic Canada. On their return many make the journey from Greenland to southwest Europe over the North Atlantic in a single non-stop flight of 2,000–3,000km.

Deserts

Most migrants flying between Europe and Africa must cross the natural barrier of the Sahara Desert to reach rich feeding grounds in central and southern Africa. The desert is 1,000–1,500km from north to south with day temperatures of 28–35°C and sand temperature of 70°C. There are oases, but while these are used by some birds, the majority make non-stop crossings on a broad front. If they do stop they usually simply rest during the day and continue their flight at night.

Many migrants from Africa, such as Redstarts and Yellow Wagtails, are becoming scarcer in Britain and Europe in spring, and while the cause is not clear, the expanding desert due to human activity in both the northern and southern fringes of the Sahara, plus the effects of climate change increasing desertification, is making this crossing longer and even more challenging for the birds.

HOW HIGH TO FLY

Some migrants, such as Swallows, fly close to the ground when migrating over land and may feed as they fly. But other species fly much higher and out of sight of us on the ground.

It has been discovered from radar studies that most small birds fly at 1,500m above the ground, but some are regularly flying at 3,000m, and others reach even greater heights, especially for long migrations.

For some small migrants, crossing the Mediterranean holds an unexpected danger. As they gather, waiting for the right weather conditions, there are predators waiting to take advantage of these tired and often inexperienced songbirds. Eleonora's Falcon times its breeding for late summer, when there is a ready supply of small migratory birds, which are easy meals for its own young.

High-flyers

A pilot of a light aircraft once saw swans, presumably Whooper Swans, flying from Iceland to Ireland at 8,000m and they did not appear distressed. At this height they would benefit from the greater air speed, but would have reduced oxygen. Air temperature would have been a chilly -42°C.

Not surprisingly the height chosen tends to be at an altitude with the most favourable winds. At higher altitudes speeds increase as the air thins, but the temperature drops. The birds expend considerable effort climbing to these great heights, and many short journeys are made at lower altitudes.

Bar-headed Geese breed in highland lakes of Tibet and winter in India. When migrating to their winter quarters they reach altitudes of 9,000m – they have been seen flying over the mountains of the Himalayas, including Everest itself.

DAY OR NIGHT?

Wheatears, Sedge Warblers and other species that make long non-stop flights must fly by both day and night, and may continue flying for 60 hours or more. Others, making shorter journeys, stop to feed several times before reaching their winter quarters. However, many small birds travel mostly at night, when they avoid predators, and feed mostly in the day.

Night migration becomes obvious in Britain and much of western Europe in autumn, when thrushes from northern Europe and Iceland, especially Redwings, migrate south in

loose flocks after dark. Individual Redwings make frequent
'seeep' contact calls, helping to keep flocks together in the
dark. It is one of the magical experiences of early winter to
hear, on a clear night, the calls of these long-distance migrants
streaming overhead.

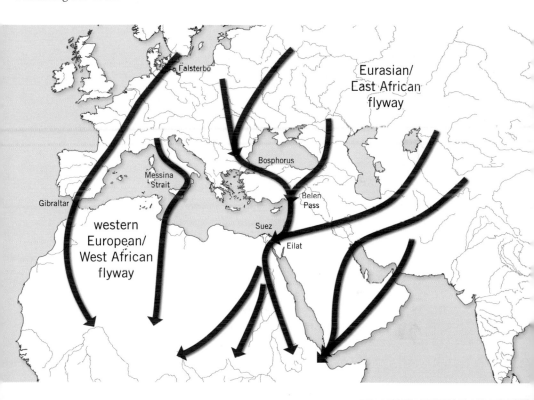

Falsterbo

Eurasian/
East African
flyway

Bosphorus

Messina
Strait

Belen
Pass

Gibraltar

Suez

western
European/
West African
flyway

Eilat

Flight paths

Major routes for soaring species have evolved which lead the
migrants to the narrowest sea crossings, or avoid the sea completely.
From Europe the three major routes are: south through Spain and
across the Straits of Gibraltar to Morocco, south through Italy and
Scilly and across the Mediterranean to Tunisia; and the major route
from eastern Europe through the Bosphorus and down the eastern
Mediterranean coast through Lebanon and Israel, and along the
Great Rift Valley into Africa.

While most birds use a flapping flight to drive themselves forwards, many larger migrants fly by day, and use warm currents of air rising off hillsides and updraughts from ridges to gain height. From the top of one such current or thermal, they can peel off and glide onwards until they find another to take them up again, and so on until the sun starts to set.

This gliding flight is almost effortless, and a very efficient way for storks, pelicans and large birds of prey to travel for hundreds of miles without expending much energy. There are, however, limitations as they have to roost in a new location each night. There are also relatively few routes from Europe to Africa that allow birds to move long distances in this way, and sea crossings become a barrier. As a result large flocks tend to merge along a small number of well-used 'flyways'.

FLYWAYS ACROSS THE WORLD

In addition to lines of hills used by soaring birds, there are other natural routes that lend themselves to successful migration.

In western Europe, food-rich estuaries provide 'stepping stones' from the Arctic to Africa. In Britain the great estuaries of the Wash, Morecambe Bay and the Solway are some of the sites where thousands of waders stop to feed and moult before travelling on to Africa. The Wadden Sea off the coast of Holland, Germany and Denmark is another area of international importance.

Many seabirds and waders will follow coastlines, even when this is not the most direct route. Other species, which travel over land, become concentrated in places that avoid the highest mountains and other natural barriers.

Some birds will travel first in one direction, but after stopping they re-orientate and fly in a different direction. Lesser Whitethroats fly southeast across Europe to the Middle East, where they need to change to a southwesterly direction to reach their African wintering grounds. Intriguingly, even birds in captivity show preferences for different directions at appropriate times, showing they have an internal clock linked

Holding the baby

Female Dotterels have brighter plumage, arrive earlier on their breeding territories and compete for males. They leave the males to incubate and tend the young and may move on to another area where they mate with another male and rear another family.

to a seasonal calendar (an endogenous rhythm).

Unless travelling directly north–south, the shortest distance between two places on the earth's surface is by a great circle route, as the earth is a sphere and not a two-dimensional map (this becomes obvious to air passengers crossing the Atlantic from Europe to North America who find themselves over Greenland!). Navigating along a great circle route would require delicate navigation, and as yet no direct evidence exists for birds using this strategy, but as satellite tracking becomes more common, who knows what more we will discover?

One surprising route is used by some of the large albatrosses. The Grey-headed Albatross has been shown by geolocation loggers to circumnavigate Antarctica, which takes it right round the earth without a major change of direction. This journey may be accomplished in as little as 46 days.

Leapfrog and displacement
One curious but quite common pattern of migration is that of one population overflying another, and flying further than

New tactics

Look in an older bird book and you will see that Blackcaps (female illustrated) are described as summer migrants to northern and western Europe, with the breeding population migrating south to the Mediterranean basin and North Africa for the winter. This is still the case, but over the last 50 years an increasing number of Blackcaps are also being seen in Britain and Ireland in winter.

At first it was thought that these had failed to migrate, but ringing recoveries show they are not British breeding birds but come here from northern and central Europe. It is assumed that the first birds that followed this 'incorrect' migratory path passed on the genetic change to their offspring, and because the new behaviour is proving successful, the numbers seen in winter continue to increase year on year. An interesting postscript to this story is the part played by garden birdwatchers, who have provided many of the reports, and may also be helping the population by providing bird food in winter.

would appear necessary. This is called 'leapfrog' migration.

Ringed Plovers breed from the high Arctic as far south as northern France. The European population may move no further than southern Europe for the winter, but the Arctic populations pass through Europe and continue onward to winter in Africa. The Arctic breeders tend to be slightly smaller birds and this arrangement prevents adverse competition for food in winter. It has also resulted in a change in the pattern of moult, with the European birds moulting quickly in summer before moving south, but Arctic birds postponing their moult until the winter.

Replacement by the same species also takes place, which helps to disguise migration. Many of the Shovelers that breed in Britain migrate to southern Europe and some reach North Africa. Other populations of Shovelers from northern Europe and Russia move southwest and many of these reach western Europe and the British Isles – leading to the false impression that the species is resident.

Recently, ringing recoveries have also shown that many of the Merlins that breed in Britain move to mainland Europe, but the British Isles receives an influx of Icelandic and Scandinavian birds, and so Merlins may be seen throughout the year.

Invasions

From time to time large numbers of migrants appear in what is sometimes called an invasion or irruption. One example is the Waxwing, which breeds in northern Europe and feeds on insects in summer. In winter it feeds almost exclusively on berries, and becomes nomadic as it follows its ever diminishing food supplies. In some years very few visit the British Isles, but in years of food shortage or an increase in the population many will move southwest and cross the North Sea in their search for berries.

Crossbills are also unpredictable in their migrations. They feed almost exclusively on the seeds of conifer trees. Cone crops are variable and in some areas in some years no cones are produced and no seeds are available. This food shortage forces the Crossbills to leave their woods and seek out new feeding areas. When new supplies are located the birds are likely to settle down and breed – sometimes starting very early in the year, with snow still on the ground. Therefore, from time to time, Crossbills from pine woods without food will irrupt and invade new areas until the food runs out and they are forced to move again. They differ from other migrants as they generally only move once in a year and there will be no return movement.

Migrating to moult

Birds that migrate need to fit their plumage moult into their annual cycle. Some species, for example Blackcaps, moult before they migrate. Others, like many of the Knots from the

Waxwings are nomadic migrants and move on to new areas when their food supplies of berries are exhausted. The speed and length of their migration varies from year to year – the winter of 2010–2011 saw one of the largest UK invasions on record. They are frequently seen around supermarkets and shopping malls where ornamental berry-bearing shrubs have been planted. Each Waxwing may eat 600–1,000 berries a day!

Arctic, suspend their moult until they arrive at a suitable stopping place where they moult and feed before continuing their journey. Some, like the Sedge Warbler, wait until they reach their winter quarters in Africa.

Some species make special migrations in order to reach moulting areas. Many wildfowl become flightless for a time while they moult and regrow their flight feathers, and in preparation for this may travel to suitable moulting areas that provide food and safety during this vulnerable period. In some populations it is juveniles and non-breeding birds that make this journey, but in the case of Shelduck, adults leave their young and make a lengthy journey – in western Europe this is mainly to the Wadden Sea off the coast of Germany.

Family differences

Wader chicks are able to feed themselves straight after hatching, and it is not unusual for a parent, often the female, to leave them. Dunlins from the Arctic have a very short breeding season and the females start their migration soon after the young have hatched. Later the males start to move south, but it may be several days later before the young birds start their migration. This means that waves of Dunlins of different ages and sexes arrive at different times along the coasts of western Europe.

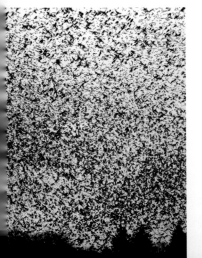

Beech holiday

European Bramblings desert their northerly breeding areas in winter. They will search for food, especially beechmast, and may visit different places in different winters. In areas where food is plentiful they sometimes gather in their hundreds or even thousands. In the winter of 1951–1952, when there was a dearth of food in Sweden and Denmark but plenty in Switzerland, an estimated 70 million birds gathered in one Swiss town and a further 30–40 million in another.

HOW DO BIRDS NAVIGATE?

The ability of birds to accurately navigate their way to suitable wintering grounds and then back to their nest-sites has fascinated both scientists and people in general for years.

There is no one simple explanation, and there remain areas of mystery. As we have already seen, there are many different migration strategies among different species with different requirements for survival.

In late summer Knot from the Arctic gather on European estuaries to feed and moult, and many will migrate onwards to African coasts for the winter. Here Knot in red breeding plumage mix with juveniles and birds that have begun to moult.

Innate behaviour

Experiments have shown that young birds have a precise sense of direction, and are therefore programmed to set out on a particular route. This is written in the genes and so is inherited. As birds become older they acquire additional skills and experience, which are added to their innate sense of both direction and time to allow fine-tuning of migratory behaviour.

Familiar landmarks can be remembered, and birds appear to have a 'map sense' allowing them to return to earlier nest-sites

Auntie Shelduck

Many adult Shelducks abandon their young and leave their nesting areas, moving off to traditional moulting areas of shallow water, which provide food and safety. A few non-breeding females sometimes continue to accompany crèches of young, and these birds have been described as 'aunties' by scientists!

or colonies, even if displaced by poor weather. Birds also have an 'optimal travel distance', whether that be covered in one marathon flight or several flights over shorter distances.

Older birds can compensate for problems more easily than young ones, as has been shown through experiments with Starlings, where birds have been removed from one place and released at another – experienced adults can still find their way home. This navigational ability has become the basis of racing 'homing pigeons'.

An onboard compass

Finding and holding a precise direction requires some form of compass. Research has found that birds respond to celestial clues. The sun is the most obvious point of reference, but an internal clock is also required to allow for its daily 'movements'.

However, many birds migrate by night. The moon's movements make it an unreliable point of reference, but there is compelling evidence that birds recognise major

star constellations and especially fixed points such as the
North Star. Experiments in planetariums have confirmed
this, although how species that pass over the equator
to the southern hemisphere adjust for the change in the
constellations is still unclear.

Giant magnet

Navigating in a set direction and using the sun by day and the
stars by night is all very well, but in overcast conditions these
clues become useless. Radar has shown birds can orientate
even under dense cloud. The additional sense that birds are
able to use is their ability to detect the earth's geomagnetic
field. In low latitudes it has been shown that this is enough
to help them determine direction, but at higher latitudes,
nearer the Arctic, this magnetic force may need to be used in
conjunction with other clues, such as sun or stars.

Subsidiary senses

There are other senses that birds use to aid navigation.
Infrasound is the noise emitted from ground level, which can

Early independence

An innate, inherited sense of
direction helps explain how a
juvenile Cuckoo finds its way to
its African wintering grounds,
when its parents have already
made the journey several weeks
earlier, and its foster parents
winter in a different area. The
young Cuckoo also flies 'an
optimum distance', which brings
it to traditional wintering areas.

Large flocks of migrating geese are composed of many separate families. Parents and their offspring of that year travel together and remain together for the winter.

be detected by flying birds. For example the sound of the ocean is different to that of land, and deserts differ from mountains – all these are audible clues that can be used by migrants to help their navigation.

Some seabirds, especially 'tubenoses' such as shearwaters and petrels, have a sense of smell, but it is uncertain how important this is in navigation. There is a strong suspicion that it may help them locate their burrows in the dark, but as yet this remains unproven.

Parental guidance

Some species make their migrations in large flocks. Flocks help inexperienced birds reach their destinations, and some species, especially geese, swans and cranes, also habitually migrate in family units, with the juveniles benefitting from the experience of their parents both on the route and in finding the best feeding and roosting areas once they arrive in their wintering areas.

FUEL FOR THE JOURNEY

Acquiring the energy to make these journeys usually needs special adaptation. Many migrants put on extra weight, especially body fat, at the time of migration with some, like Garden Warblers, more than doubling their body weight from the normal 18g to 37g before setting off to fly over the Sahara Desert.

Not all small migrants are this extreme; many will accumulate less fat, increasing their body weight by 10–30%, and make their journeys in shorter flights – often at night. They will use the daytime stop-over places to feed and replenish their 'fuel' for the next leg of their journey. Birds that make non-stop flights across large stretches of sea or desert need the largest fuel reserves.

There are pros and cons to both strategies. Longer flights require most energy and take birds across the most inhospitable habitats – potentially disastrous if they need to land. Shorter flights require finding several suitable stopping places with enough food to replace lost fat, and when resting, tired migrants are easy prey for predators.

Geese that nest in the Arctic may carry more fat than they need for their migration; the reserve is for the harsh conditions they may encounter and will allow them to survive egg-laying and incubation at times when food might be scarce.

Birds may change their diet to gain the additional weight necessary for migration – Sedge Warblers start eating aphids a couple of weeks before migrating – and there may also be metabolic changes, and increases in the size of flight muscles

Before migration the Sedge Warbler changes its diet and puts on extra weight to fuel its journey. In about two weeks the body weight almost doubles. It has been calculated that these reserves would allow a Sedge Warbler to make a non-stop flight across much of Europe, the Mediterranean and the Sahara Desert to reach its African wintering areas without stopping.

for the migration season. It has been observed that migrant birds kept in captivity will also increase body weight around the normal time of their migration, and lose it again without making any substantial flights.

NO WINGS?

The fact that penguins can't fly does not prevent them migrating. King Penguins will walk for miles over featureless sea-ice to return to their nesting colonies. The much smaller Magellanic Penguin swims northwards up the coast of South America after the breeding season, a journey of 2,700km. Some of this journey will be made under water, much easier going for the bird.

Some other species will walk their migrations. The Blue Grouse of North America is capable of flight but appears to cover much of its migration on foot. Emus in Australia are flightless, but also make seasonal migrations.

VAGRANCY

From time to time migrants lose their way and appear in places where they are not expected. The hobby of 'twitching' involves trying to see as many species of birds as possible, and it is these lost 'vagrants', perhaps the first of their species to turn up in a particular area, that attract the largest numbers of twitchers.

Some of these vagrants are birds that have overshot their target. Hoopoes and White Storks which winter in Africa and breed mainly in Southern Europe sometimes appear in southern England in spring.

Most migration is influenced by clear skies and following winds, but stronger side winds, especially at night, can push birds off course. Adults find it easier to re-orientate themselves than juveniles. Birds away from their normal migration routes will land, often near the coast, to recover before trying to continue their journeys.

Yellow-browed Warblers from the foothills of the Himalayas and Pallas's Warblers which breed in Russia and parts of

This exotic-looking Bee-eater breeds mainly around the Mediterranean. In some springs, individuals overshoot mainland Europe and reach the British Isles, and occasionally even nest here.

Aliens abroad

What becomes of these lost birds? For many small birds that have strayed long distances, the chances of successfully recovering from the journey, let alone resuming their correct migratory course, are very slim, and they will probably perish. Larger and more robust birds have a better chance of survival. A few individuals may settle permanently in their new country, even fraternising with the locals, and adopt a new annual migratory path. In recent years, some Ring-billed Gulls, originally from North America, have been spending their winters in the Essex town of Westcliff-on-Sea.

Asia both normally winter in South-east Asia. However, each year a few juveniles, for reasons that are unclear, migrate in the opposite direction, and their fat reserves bring them to a relatively small area of the British east coast.

Other birds migrating southwards from North America sometimes get swept out into the Atlantic, and some of the survivors reach the British Isles, especially western headlands and islands like the Scilly Isles, in autumn.

IS IT ALL WORTH IT?

The casualties from migration are enormous, especially among inexperienced juveniles (for the Black-throated Blue Warbler of North America, it is estimated that 85% of the annual mortality was during the two-way migratory flight). However, as migration has evolved through millions of years, this loss has been more than compensated for by the benefits of travelling to exploit safe breeding areas and food-rich places to rear young, and to return to suitable winter quarters where food is plentiful. As climate changes and food supplies alter, so birds' migrations will need to change in order for the species to survive – it has always been so.

Feeding

Many people get pleasure from watching birds feed, whether it is by providing bird feeders in the garden or visiting a park to feed the ducks. We can also marvel at Gannets diving for fish by plunging headlong into the sea, or smile at a Puffin with a bill full of sand eels. In fact, there is almost nothing that grows or moves on the surface of the earth that birds have not exploited in some way, and their food requirements have helped shape the way they have adapted into the wonderful variety of species that share our planet today.

DIETARY NICHES

Survival depends on finding enough food to stay alive; too much competition and individuals will starve, but feeding in flocks can also save time and energy in the daily search for

Puffins are unusual in being able to carry several fish at once.

food. Species have evolved with different requirements and, in general, no two related species living in the same habitat have exactly the same food requirements or foraging strategies.

Northern European finches, for example, all have the conical bills of typical seed-eaters, but closer study reveals that they have either evolved to live in different habitats, or specialise in different foods.

The Twite and the Linnet, both finches, share many food preferences, but live in different habitats. The Chaffinch and the Brambling have much in common, but the Brambling tends to be more northerly and have a stronger bill with which it can open beechmast. Hawfinches crack open larger stronger seeds and Bullfinches specialise in fruit buds.

Greenfinches take larger seeds than Goldfinches; the latter also have longer bills for extracting seeds from thistles, while the closely related Siskin has a stronger bill for eating alder seeds. Obviously, there is also overlap in their food choices – especially when there is an abundance of easily obtainable food.

While all crossbills feed on the seeds from pine cones, the different species of crossbill usually prefer cones from particular species of conifer.

FINDING FOOD

Most birds find food by sight. A Hen Harrier will quarter its moorland territory methodically, searching the ground for prey. Blackbirds feeding on the ground tend to hop along a zigzag course when searching for their food, but will change course once a prey item is discovered and will then concentrate on the new area where food has been found.

A Kestrel hovers over a potential feeding site or sits patiently on a nearby perch watching for movements of insects or small mammals. Besides its keen eyesight it has another advantage – the ability to see ultraviolet light. Vole urine reflects ultraviolet light, and this enables the Kestrel to see the 'runs' along which the voles move through long grass.

Some birds listen for their prey. A thrush cocking its head on one side is probably looking as well as listening, but a Blackbird can hear the movement of invertebrates in leaf litter and uses this to locate its prey. Woodpeckers probably hear the sounds of insect larvae moving within wood.

For nocturnal owls, hearing is essential for finding prey. With a unique arrangement of asymmetrical ears (one higher on the skull than the other) a Barn Owl is able to accurately pinpoint sound – even in total darkness.

Waders with long bills, such as the Snipe, probe into mud and silt and feel their prey with nerve endings at the tips of their bills. Plovers take most food from the surface and feed more by sight; they have the benefit of large eyes for continuing to feed after dark.

WHEN TO FEED?

Small birds spend most of their day searching for food. Blue Tits may spend 85% of daylight hours in a hunt for food during winter, and a similar amount of time foraging when feeding their young. The smaller Coal Tit may exceed this with 90% of daylight hours spent foraging.

Most birds feed by day and rest by night, but for waders and

Ordered pecking

Redshanks feed energetically, but how much of this pecking is successful? Ornithologists studying waders look at the pecking rate and also at the number of successful strikes. One survey of Redshanks found them to peck 60–71 times a minute and they were successful 70–90% of the time, the rate depending on the season and the density of the food.

Female Eiders incubate their eggs alone, with no help from a mate, and during this time they will not feed. Instead they rely on stored fat. If they leave the nest it will only be for a short time in order to drink.

ducks, which feed around our estuaries, the ebb and flow of the tide is more critical. If the conditions are right they will continue feeding after dark. The highest tides, whether during the day or night, drive Knots, Dunlins and other waders off their feeding grounds into high-tide roosts until the mudflats are revealed again.

Some seasonal changes in food are linked to the birds' annual cycle. Before departing on their annual migration, Sedge Warblers spend more time feeding, and eat a lot of the aphids that are found in reedbeds at that season – this helps build the reserves of fat required for migration.

Some food is more available at particular times of day. 'The early bird catches the worm' may be a familiar proverb, but it is based on the fact that worms are more available to birds such as thrushes in the moist dew of the early morning.

Birds are also opportunistic and will take advantage of new seasonal food supplies. One example can be seen on days in late summer when ants swarm. Looking up into the sky, you could see a variety of species exploiting this short-lived bonanza, from Starlings to Black-headed Gulls.

It seems obvious that birds feed when hungry, but so much of their behaviour is geared towards feeding that sometimes they don't seem to know when to stop, and will continue to feed after their appetite has been satiated. At this point they discard food or store it for later.

For much of the year jays are secretive woodland birds, more often heard than seen. In autumn many are seen in more open countryside, as they move from their woods in search of acorns on isolated oak trees.

Food storing

Several species will put food aside in a safe place for later consumption. It is an efficient method of dealing with a short-term glut. Jays in woods store acorns; these will be carefully hidden, usually buried, and retrieved again at a later date. Some are forgotten, and this habit has contributed to the spread of oak woods.

Coal Tits in gardens will often hide food, especially sunflower seeds. This behaviour is sometimes obvious as they come and go rapidly from a bird feeder, while other tits will take more time and eat on or near the feeding site. Much of this stored food is eaten quite quickly and is not generally kept for long-term storage.

Great Spotted Woodpeckers store pine cones – sometimes wedging them in a crevice in a tree trunk. In North America, groups of up to a dozen Acorn Woodpeckers excavate special holes that are then filled with acorns. The group co-operates to

Shrikes like this Red-backed catch large insects, small birds and mammals. They often store their prey on thorn bushes, colloquially known as 'larders'.

defend this 'larder' against intruders.

Nutcrackers store hazelnuts in autumn for feeding young in spring. Their ability to recover the hoards (usually 15–20 nuts in each hoard) is quite remarkable, and one piece of research identified an 86% success rate. Some birds were even able to recover their hoard when the ground was covered with snow!

Birds of prey such as the Kestrel sometimes cache food that is not immediately required. While not related, shrikes share some of the same characteristics as small birds of prey and also hoard their prey. Their food items are impaled on thorns, or sometimes on barbed wire fences, and this habit has earned shrikes the vernacular name of 'butcher birds'.

WHAT TO EAT

Some birds are specialists, concentrating on a particular food type, such as seeds from pine cones in the case of crossbills. Others, such as gulls, are omnivorous and feed on a wide variety of food from carrion to insects and fruit – and they also rob other birds of their food.

It's not unusual for a species to have a different diet from one time of year to another. Redwings feed on invertebrates during the summer, but become fruit eaters in autumn – taking advantage of the huge crop of berries in the countryside to

Redwings and Fieldfares feed on insects in summer, berries in autumn and worms and other invertebrates when berries run out in winter.

Mistletoe berries are popular with birds – especially Mistle Thrushes. The plant's seeds, wrapped in a sticky gel which survives digestion, are taken from tree to tree by birds; they gain a hold on a new plant when wiped from a bird's bill or ejected from the body as a dropping.

which they migrate. When the berries run out they can move to fields to hunt worms and other invertebrates found on areas of short turf.

Another winter visitor to Britain, the Waxwing, feeds on insects in summer – many of them captured in flight. In winter, though, it feeds exclusively on berries, and thus becomes a nomadic migrant on a tireless search for these fruits. It is well equipped, with a wider gape than thrushes of a similar size, which is helpful for swallowing berries, and also has a large liver – typical of fruit-eating birds.

A surprising number of birds that we might consider seed eaters, such as Goldfinches, Reed Buntings and Coal Tits, are dependent on insects for food in summer – indeed, insects are essential food for their young. Gamebirds such as Grey Partridge are mainly vegetarian, but their young are almost exclusively insectivorous, especially liking the larvae of sawflies. The use of pesticides against insects on crops is one of the causes of the Grey Partridge's decline in agricultural areas.

Exceptional adaptations

Within some bird families, there are particular species which have evolved to pursue a different dietary path to their close relatives. Woodpeckers primarily feed on invertebrates living in trees, where they can use their long tongues to probe into crevices and holes, and extract the insects and their larvae. However, Green Woodpeckers have further adapted to feed on the ground, and use their tongues to extract ants and their larvae from underground nests. Sap-suckers in North America are also woodpeckers and make small holes from which they take the sap from the living tree.

Kingfishers in Europe and the Americas are perfectly adapted to feed on their prey, mainly fish, taken from the water by diving. In other part of the world members of the kingfisher family feed on other foods, often invertebrates, while the Laughing Kookaburra, a large kingfisher living in Australia, feeds on lizards, small rodents and even snakes.

Sexual differences

In a few cases, what a bird eats depends on its sex. In some birds of prey, particularly bird-hunting species like the Sparrowhawk, there is a marked size difference between male and female. A large female can weigh almost twice as much as a small male. The size difference reduces direct competition for food between male and female, and means a greater range of prey is available for the family. During incubation and while the chicks are small, the male provides most of the food, leaving the female to incubate and defend the nest. Later the female joins in the hunting and helps feed the growing chicks with larger prey.

TECHNIQUES FOR FEEDING

Many feeding techniques are characteristic of a particular species. We're all familiar with the low swooping and twisting flight of a Swallow chasing flies over a summer field.

Two members of the thrush family have developed individual feeding habitats. Blackbirds can be seen (and heard) turning over dead leaves under shrubs in their search for invertebrates. Song Thrushes have evolved a unique way of opening snail shells. They hold the shell by its lip and smack it against a stone or other hard object, traditionally called 'a thrush's anvil'.

Sparrowhawks have a greater size difference between male and female than any other bird of prey, and females may weigh almost twice as much as males – they also generally live longer.

Coughing up

Most birds of prey tear up their prey, but owls swallow it whole. The indigestible parts are regurgitated as pellets. Typical pellets contain fur and bones, from which the prey items can be identified. Small mammal skulls are often complete and recognisable. This helps provide a picture of the owl's hunting success, as well as a valuable cross-section of the small mammals living in a particular area. Some other birds, such as gulls, waders, crows and thrushes. also produce pellets of hard ingestible prey parts.

Shellfish present a similar problem to some coastal birds. Oystercatchers have evolved two methods for opening bivalves such as cockles and mussels. If the two halves of the shell are already slightly parted they insert the tip of their bill and prise them fully open. If the shell is closed, then a series of sharp blows with the bill will usually gain entry. This method of feeding is not possible for young Oystercatchers and so adults feed their chicks until they are able to fly, which is unusual as most young waders feed themselves.

Gulls and crows cannot use their bills to open shellfish, but have learnt to fly up and drop a mussel from a reasonable height. Provided it hits a hard object such as a rock or harbour wall, the shell is likely to smash and reveal the contents. In places where this happens frequently, Turnstones have learnt that the sound of dropping shells means food if they can arrive quickly enough and claim the prize before the gull can descend to retrieve its prey.

Food in trees

A single tree can offer a varied feast for many different bird species. While some such as pigeons and parrots eat leaves, and Bullfinches take buds, most birds are more interested in the fruits, or invertebrates that feed on the wood and foliage.

As we have seen, woodpeckers search out creatures living
on or under the bark. Tits will feed on leaf-eating caterpillars.
Blue Tits are more likely to be nearer the tops of trees and
Great Tits lower down or on the ground. In autumn and winter
they will also be hunting for tree seeds – and their ability
to hang upside down to reach seeds at the ends of the most
delicate branches is an essential skill.

Nuthatches have taken seed- and nut-eating further. They
habitually search out large seeds and small nuts and wedge
them in crevasses in the tree bark, and hack them open – thus
acquiring their well-deserved English name.

Food in water

There are three main ways birds exploit food in the water. They
may feed round the water's shore, dabble from the surface
while swimming or dive below the surface, either from the air
or while swimming. This holds true for fresh water, including
rivers, and for the sea.

Life on the edge · Herons and egrets hunt by stealth around
the water's edge. They stand for minutes like statues before
striking out at their prey; whether fish, frog or small mammal.
Some take the initiative and stir up the water, or spread their
wings, either to help them see their prey better or encourage
fish to swim into the shade. Placing bait on the water to attract
fish has been observed in several species of smaller herons.

Dabbling ducks · The ducks which dabble, including Mallard,
Teal and Shoveler, mainly feed from the surface. Sometimes
they also upend, to reach food from deeper water. They obtain
food by pumping water or mud with their tongue through the
small serrations (called lamellae) along the edges of their bill.
This action filters out tiny invertebrates that they then eat.
Dabbling ducks avoid competition by feeding in different zones
and on different food.

Diving · Many birds obtain their food by diving into water.

Shovelers sometimes feed in
groups and benefit from food
stirred up by others. This may
result in them feeding in a line or
even in a circle, as each benefits
from the action of the feet of the
next bird along.

Some ducks, such as Pochard, are mainly vegetarian and dive to find pondweed, stonewort and other water plants growing below the water surface. Tufted Ducks and Eiders feed on mussels that live at the bottom of lakes or under the sea.

Most other diving birds are fish hunters. They include the 'sawbill' ducks, such as Goosander and the mergansers, so called because of their pronounced lamellae that help them to grip a slippery fish.

Among the seabirds, Guillemots have exceptionally good eyesight under water and some Black Guillemots have been seen feeding after dark. A study of Brünnich's Guillemots showed that they could reach a depth of 135m and stay submerged for over three minutes, although an average dive was 64m deep and lasted 55 seconds. The big penguins, King and Emperor, regularly dive to 200m or more, and Emperors may stay under water for more than 10 minutes.

Plunge-diving is a dramatic way to capture fish. Around their colonies, Gannets can be observed making spectacular headlong dives en masse. When fishing they generally fly at 9–15m with bill pointing down, and then drop with wings pulled back, sometimes straight and sometimes angled, hitting the water like a spear at 100km/h (60mph). This dive will take them about 3m below the surface where they swim for only a few seconds while they hunt their fish.

Quite a different plunge-dive is that performed by the Osprey. It also searches for fish from the air before dropping into a dive. Unlike the Gannet, it enters the water feet first and grasps its prey with its talons, causing a huge splash. It may almost completely disappear below the water. If successful it rises again with a fish gripped in its talons. The prey is carried 'torpedo-fashion', aligned with the bird's body, to minimise wind resistance.

Most passerines (perching birds) are strictly land birds, but one group is truly aquatic. The dippers are rotund, rather thrush-like birds which dive in fast-flowing water and stay submerged for several seconds while they search for prey.

Dippers will feed on land, but much of their insect food is

Diving ducks dive to feed on stonewort and other water plants. They have skulls with few air spaces, which helps reduce buoyancy and problems of changing pressure under water.

obtained from the beds of fast-flowing rivers and streams. To reach this underwater larder the bird frequently perches on a stone in the water and submerges its head to see below the surface. Then it will sometimes swim, or more often walk, into the water and submerge itself as it searches for food. Under water it uses its wings to maintain its position and for propulsion, and it can even travel upstream against the current, clinging to stones on the riverbed as it moves along.

Aerial feeders

Probably the best known aerial feeders are the swallows and swifts. For these agile birds, flight isn't just a means of getting from A to B. These habitually feed 'on the wing', catching flying insects in their wide gapes. The gape is lined with bristles, which were once thought to improve the catching ability, but

Ospreys are especially well-adapted to catching fish. Their talons are curved inwards to improve grip, and the fourth toe is 'reversible', meaning it can grip from the opposite direction. The pads of the feet have sharp spines that help grasp the prey. This fisher par excellence also has nostrils that can be closed at the moment of impact with the water.

are now considered to protect the bird's eyes.

Many other birds also catch insects in the air. Chaffinches frequently launch themselves from their perches to grasp a passing insect. However, the masters of this technique are the flycatchers. A Spotted Flycatcher will dart out from its perch in a graceful sweep to snap at a passing insect and return to the same perch – listen carefully and you may hear the snap of its bill as it catches the insect.

Aerial hunters sometimes become the hunted. Falcons have perfected the art of catching smaller birds in flight. The largest, the Peregrine, has several methods. It may arrive fast, at low level, and take a bird by surprise; but more dramatically it will stoop on its prey from above, in a dive claimed to be the fastest of any bird. Measurement is difficult, but one authority claims the speed at which it strikes its prey is likely to be 120–150mph.

The Hobby is a smaller relative of the Peregrine and is a fast and dramatic aerial hunter, which also sometimes hovers, rather like a Kestrel. The Hobby feeds on large insects seized in flight, particularly dragonflies – which are themselves fast and agile fliers. Hobbies also regularly capture Swallows and martins and even take Swifts.

Social feeding

If food is plentiful there can be benefits in feeding in flocks. Flocks mean more birds to join the search and more eyes to look out for approaching danger.

Some species will join others as they feed. Mute Swans are wasteful feeders. As they pull waterweed up from deep water some of it floats away. Other waterfowl will make use of the newly exposed food, a habit that is called commensal feeding. The other waterfowl may be Mallard, Gadwall and even Pochard and Coot. The last two species dive for their food, but will happily accept a free meal on the surface.

Golden Plovers breed in uplands, often close to, or within, breeding territories of the much smaller Dunlin. The feeding habits of the two birds could not be more different: with

the Dunlin feeding with its head down, like a little sewing machine, and the Golden Plover standing tall and erect, stepping forwards, bending quickly to snatch food and then standing erect again. The plover is on constant lookout for danger and gives security to the little Dunlin, which feeds more efficiently in company with the larger bird. If alarmed, they are very likely to fly off together and resume their partnership elsewhere. This curious habit has earned the Dunlin the local name of 'Plovers' Page' in northern England and Scotland, and 'Plovers' Slave' in Iceland.

In winter, flocks of Golden Plovers are attracted to flocks of feeding Lapwings. Lapwings feed on earthworms and gather where food is most abundant. Golden Plovers joining them benefit from the food supply, without wasting time searching for their own separate feeding areas.

Mute Swans uproot and dislodge three times the vegetation that they actually ingest, but their inefficient habits aid other waterfowl's foraging as they stir up vegetation deep in the water that smaller waterfowl, including ducks like this Gadwall, cannot otherwise reach.

Pirates

Those Golden Plover and Lapwing flocks don't have things all their own way. They often attract the attention of Black-headed Gulls, which harass them and try to rob them of their food. This thievery is called kleptoparasitism, and it is taken to the highest level by the skua family. Skuas specialise in robbing other birds of their food. Most skuas are coastal nesters, and harass seabirds such as terns, Guillemots, Puffins and even Gannets for their fish, targeting individuals on their way back to their nesting colonies with food for their young. Skuas migrate to the same winter areas as terns, where they continue to rob the smaller birds.

Kleptoparasitism or piracy is also the primary feeding technique of the Magnificent Frigatebird, a spectacular seabird found in tropical and sub-tropical waters. Large and menacing with a vicious-looking bill, it is very adept at making smaller birds drop their catches.

Skuas are aerial masters, expertly twisting and turning to follow every manoeuvre of their victim, until the food is eventually dropped, and then caught or picked up by the skua. Here an Arctic Skua is pirating an Arctic Tern.

Pelicans often hunt for food co-operatively by swimming in a line or semi-circle.

Synchronised swimming

Several species co-operate in groups and hunt communally. Goosanders will sometimes swim in lines or crescents in the water as they drive fish in front of them. Then at the same time they will dive and seize their prey. Pelicans have a similar strategy but swim with their bills under water, using the pouches as fishing nets.

TOOLS FOR THE JOB

Birds' bills have evolved to meet their feeding requirements. Some are multi-purpose and are a clue to an omnivorous diet. Crows and gulls could be said to have general-purpose bills, suitable for many different purposes, and strong enough for scavenging, eating carrion and even killing small mammals or other birds.

Finches are classic seed-eaters and, in addition to a stout conical bill, have a powerful skull and large jaw muscles. They also have a groove on each side of the palate, which (aided by the bird's tongue) helps them remove the husk from the seed.

Some finches' bills are remarkable in their specialisation. Crossbills literally have a crossed tip to their bills, a unique tool for first shredding the tough scales of pine cones and then, with the help of their rather long tongue, removing the seeds hidden beneath the scale. They also have the benefit of strong feet for holding the cone while feeding.

Hawfinches have massive bills, an adaptation for cracking

Godwits use their long bills to probe deep into silt in estuaries – making use of a rich feeding zone unavailable to birds with shorter legs and bills.

open the hardest seeds. Cherry and olive stones both feature in their diet. Young Hawfinches feed on softer food until their first winter, by which time their skulls will have ossified, with two hard knobs formed inside each mandible. These knobs allow the bird to hold the seed and exert a maximum pressure that, in an experiment, was found to be the equivalent of 60–90 pounds of force – not bad for a bird weighing only 35 grams!

Puffins have spines on the roofs of their mouths and a fleshy tongue – an unusual combination that allows them to carry multiple fish at one time. Other auks also sometimes carry more than one fish, but Puffins are the masters. On one occasion a bird was observed carrying 62 fish, but five is more usual!

The long and the short

Waders, known as shorebirds in North America, have the most diverse bills of any group of birds – varying from the short and stubby to the ridiculously long, but all have a purpose. Many of these birds have long legs, but again there is variety.

Plovers, including Lapwings, have rather short bills and medium-length legs that are suited to walking or running rather than wading. They do not probe far into the soil or mud to obtain their prey. Lapwings undertake additional winter migrations if the topsoil freezes and prey goes too deep to reach.

Curlews are large waders with long, down-curved bills for probing in wet soil and mud, and long legs for walking and wading. Their curved bills help find prey by coming into contact with a greater area than a straight bill when probing in silt. It can also be used to snatch invertebrates from the surface.

Godwits have long legs for wading and long straight bills for probing for food items in the mud. The Snipe and Woodcock also have long bills, but short legs and large feet for spreading their body weight. The Woodcock is a bird of damp woodlands, while the Snipe prefers marshes and lake margins. Like other long-billed waders, the Snipe probes mud with a closed bill, but can open the tip to grasp prey while underground, through a process called rhynchokinesis.

Avocets have up-swept bills and long legs – they habitually wade and sweep their bills from side to side as they take crustaceans from brackish water.

Tongue-twisters

Woodpeckers can reach deep inside the chambers of their insect prey by using their long, extendible tongues, which are coiled inside the bird's skull, and linked to special bones and muscles. The tongue has either a sticky or barbed tip to secure the prey. In some species the tongue can project 13cm further than the tip of the bill.

Turnstones have short legs and an all-purpose stubby bill with which they will literally turn over small stones and seaweed in their search for invertebrates. They also scavenge, and will eat carrion – one observer famously observed a flock feeding on a human corpse.

Oystercatchers' bills are, as we have seen, used for stabbing or prising open shellfish. The bills often become blunted by continual stabbing but, unlike other waders' bills, it appears that they also continue to regrow.

ENERGY REQUIREMENTS

The purpose of feeding is obviously to gain the energy for a bird to survive, but just how much energy does a bird require? This can be hard to investigate. Peregrines eat meat and gain a high calorific value from their food, but they require energy to hunt, and their feeding rate can be influenced by the availability of the prey and the weather conditions. In captivity it has been estimated that a female has a metabolic rate of 72 kilocalories and requires 51.4g of meat a day to survive, but to fly and hunt she will probably require three times that amount.

Research showed that Dunlins in the Arctic needed 80 kilocalories before migrating, whereas a Dunlin wintering in Britain only requires 38.5 kilocalories to survive. For many species, preparation for migration means prolonged bouts of feeding and an activation of particular enzyme systems, leading to the storage of additional body fat (as much as a 50% increase in weight for some small species such as Pied Flycatchers). Each gramme of fat provides 9.2 kilocalories – the most efficient fuel for the long migratory flights.

Berries are much easier to find and numerous at some seasons, but their calorific value is low. A Waxwing eating 600 berries in a day will gain 43 kilocalories – only just enough to stay alive – as the daily requirement is 30–40 kilocalories.

LEARNING NEW TRICKS

Feeding habits have evolved over time, but learned feeding behaviour can spread through a population of birds

The bird that got the cream

In the 20th century, foil-topped milk bottles delivered to British and Irish doorsteps proved tempting to inquisitive Blue and Great Tits. They discovered cream under the caps, and soon tits were removing milk bottle tops across the length and breadth of the country. The habit has now died out, as fewer bottles are delivered, and fewer people opt for full-cream milk.

surprisingly quickly, as others observe the individuals who first show the new behaviour.

Both Goldfinches and Long-tailed Tits have recently started visiting gardens in larger numbers where they have discovered bird feeders and, in the case of Goldfinch, it appears the increasingly common supply of nyjer seed has prompted regular visits to feeders across Britain.

Other learned behaviour is known from birds from elsewhere in the world. In Africa, Egyptian Vultures drop bones to smash them on rocks, enabling them to get at the marrow inside. The Woodpecker Finch of the Galapagos Islands uses a thorn to extract insects from their chambers in tree trunks, as it does not have the benefit of the long tongue of a true woodpecker.

Roosting

Knot form huge flocks that wheel and turn in flight before heading to their evening roost. The largest numbers are found in The Wash, where they create a memorable natural sight not to be missed.

On winter evenings I often stand on the shores of The Wash, that great English estuary, and watch the sun sink behind Boston Stump, as the waders in front of me retreat before the flow of an incoming tide. Once the sun has gone I hear the distant sound of geese. Not just a few calls, but a wonderful musical cacophony getting closer and closer, until overhead, and against the darkening sky, squadrons of Pink-footed Geese appear; line after line after line, heading from their daytime feeding grounds on the sugarbeet fields to spend the night in the wilderness of The Wash. They arrive gracefully, gliding in, until the last moment when many stall or slideslip in the air, an action called 'wiffling', and land on the estuary mud where they will be safe from predators.

When some birds roost they tuck their heads into their back feathers.

WHAT IS A ROOST?

Roosting behaviour isn't as predictable as you might think. On moonlit nights, these Pink-footed Geese may stay out in their fields and continue feeding. On other nights they fly to somewhere where they can roost on water, usually a coastal lake or a sheltered bay.

The word roost does not really mean sleep. It is a term used for the place birds go to in order to rest, where they may or may not sleep. If you come across a Tawny Owl roosting during the day, it may be awake and turn to look at you, but it will probably not move, and you would say the bird is 'in its roost'.

A roost needs to provide two essential elements: safety from predators, and protection from the elements.

Once Starlings swarmed in their thousands on ledges of buildings and in branches of the trees of Leicester Square in the heart of London. They would squabble and chatter until eventually settling down for the night – this was their roost. Central London provided a temperature higher than that of the surrounding countryside, and they felt safe on their ledges, packed close to many others.

Starlings also have daytime roosts, which are more obvious outside the breeding season. In these roosts the birds will sing,

Inside the roosts, Starlings squabble over the best perches, with the strongest males usually winning the best positions.

preen, sleep and some just sit. From these roosts individuals will leave to forage, and often join other groups of feeding Starlings before rejoining their roosts. In the evening these small flocks join the larger night-time roosts.

Night-time Starling roosts are particularly spectacular, often with thousands of birds joining together. These have been well studied by Chris Feare and he identified four different behaviour phases: pre-roosting assemblies, entry to the roost, behaviour in the roost, and leaving the roost. The different behaviours are stereotyped and consistent from roost to roost.

Pre-roosting assemblies draw in Starlings from a wide area, usually from several miles away – up to 24 miles has been recorded in the UK and even further in North America. Groups of Starlings arrive from all directions and gather on perches or feed together in large flocks. Many preen and bathe, and flocks will sometime bathe communally together. The location of these pre-roosts may change from time to time.

Starlings then fly from their pre-roosts to enter their roosts. There tend to be two ways of doing this. Either directly, with flocks swooping straight into the main roost, or with breath-taking aerial displays, where birds mass together and twist and turn, rise and fall, separate and come together before tumbling en masse into bushes and trees. The reason for this air show is obscure but it certainly helps to draw in all the Starlings in the area to share the roost site.

In the roost the birds jostle for position, and at first there is a lot of singing and some fighting for good perches. It appears larger males take the best positions in the centre of the roost with juveniles and females on the periphery.

Singing also heralds the morning departure, as birds move up the branches and leave in waves. The singing stops before the departure of each wave, but as groups leave the singing re-starts. Radar screens shows that on calm mornings the departures are in all directions and appear like circular ripples from a pebble dropped into a pool – with a short gap before the next ripple.

In contrast, the breeding season commences for Starlings with the males leaving the communal roosts and returning to their territories, where they may roost in their nest holes.

Loafing

This is a term which can be confused with roosting, but really describes birds spending time doing nothing in particular rather than sleeping – although some may sleep as well! Pochards do much of their feeding at night, and during the day they spend a lot of time in a group swimming, not feeding and not really sleeping – although some may take naps from time to time. Gulls also spend much time loafing.

WHEN TO ROOST?

Birds roost when they are not feeding or singing or defending their territories, and usually this is at night, but obviously not for nocturnal species, such as most owls and Nightjar.

Many waders' feeding times are regulated by the tides. Species such as Bar-tailed Godwit and Grey Plover will feed at low tide, and roost when their feeding grounds are covered by

Patterns formed by Starlings leaving their roost as shown by radar.

In winter, Grey Plovers move south from their Arctic breeding grounds and many winter in estuaries in Britain and Europe. They feed on invertebrates in the mud and sand, and can only reach their food when the tide is low. Often they roost at high tide during the day and feed at low tide after dark.

the high tide. Knots have spectacular movements as they move en masse ahead of the rising or falling tide, and will roost in huge tight groups above the high-tide mark.

In the Arctic, with 24 hours of daylight in summer, seabird colonies at midnight are still full of noise and activity. Here feeding and breeding behaviour does not completely stop at any point and individuals just rest sporadically, either on their cliffs or in the water, depending on the species. Studies have shown there to be less activity and more birds resting around midnight, but it is not clear how many of these are sleeping.

Street lights and bright security lights can also have an effect on birds' daily cycles. Some, such as Robins, regularly sing where lights are bright. The lights are also known to prolong feeding times and thus reduce the time birds would be roosting. House Sparrows may roost on light fittings and benefit from the additional warmth.

SLEEP

For a short period in spring, male Nightingales sing by day and at night, and they also feed during daylight. It would seem they have short rest periods between bouts of song, but they obviously do not take long periods of what we think of as sleep.

It is thought that under normal conditions birds spend a third of each day resting, loafing or sleeping. This depends a lot on the season and the length of daylight available. Eiders at lower latitudes in Norway may spend several hours a day asleep and also sleep at night, whereas those in the Arctic may only sleep for two or three hours in total.

Watching a sleeping bird is not easy and laboratory conditions do not properly imitate those of the wild. It appears that during rest periods birds' eyes will often close, but not for long; a few seconds to two or three minutes appears normal. When sleeping a bird may turn its head and rest the bill on or under the scapular feathers on the back (the head is never actually under its wing though!)

Nightingales sing by day and by night. It is not clear when they sleep, but they may rest between bursts of song.

Keeping on a perch

So why don't birds fall off their perches when they relax and sleep? This puzzled me for years! The answer is in the arrangement of their leg muscles and tendons. As the bird lowers its body and relaxes, the 'flexor tendons' which run down the legs to the toes are automatically pulled tight, and the more the bird sinks down the tighter they become – so the grip on the branch increases the more the bird relaxes.

Most roosting birds perch across a branch so they can grip it tightly with their toes. Nightjars usually perch along a branch, which allows greater camouflage.

It seems very likely that young Swifts that leave their nests in Britain will not land again until they are ready to breed in their third or fourth year. Swifts have occasionally been seen roosting on slender branches and on the outside of buildings, but these seem to be exceptional occurrences and most non-nesting Swifts are aerial during the hours of darkness.

Roosting in flight

This may seem a contradiction in terms, but Swifts and some other aerial species roost on the wing. The first recorded observation was by a French pilot during the First World War, who found himself among a group of almost motionless Swifts at a height of about 3,300m. He even claimed to have captured two so they could be identified.

More recently there have been both radar and observer sightings of groups of almost stationary Swifts, apparently sleeping or resting at 100m or more above the ground. The birds appear to maintain their positions by facing the wind and making gentle flaps with their wings.

This behaviour is not limited to Swifts, and it seems likely many House Martins also roost on the wing. On a clear evening in late summer it is possible to hear flocks of martins calling from high above their colonies as they gather together – many of these may be young or non-breeding birds. Gradually they go higher and higher until they pass out of human vision and there, presumably, they will stay until morning.

WHERE TO ROOST

Many species roost in a nest before eggs are laid. This is especially true in hole-nesting species such as Starling and woodpeckers. Some nests continue to be used as roosts after the young have flown and, sometimes, birds will use other species' nests. Wrens will use a House Martin's nest in winter – especially in cold weather – and occasionally several Wrens will gather in a single nest.

A few species, including House Sparrow and Wren, sometimes build special autumn nests just to use as roosts. Wren roosting nests may also be used as a roost by the males in summer, while the females are incubating.

Nestboxes may be used for roosting in during the winter. We see on page 199 how Wrens have been known to use them in cold weather, but Blue Tits and Great Tits will also use them on occasions, although it seems that most tit roosts are in much smaller cavities, where birds can more easily keep warm.

Treecreepers roost individually in crevices, usually in hollows in tree bark. They nestle in, exposing the minimum amount of their bodies to the elements. One particularly favoured tree is the Giant Sequoia or Wellingtonia, which is native to North America but has been widely planted in parks and large gardens in Europe. The bark of this tree is unusually soft (like spongy cork) and Treecreepers are able to fashion an elliptical hollow in the bark. In daytime these little hollows are clearly visible on the sheltered side of a tree and it is possible to discover if they are used by looking for a tell-tale white dropping below the hollow.

Many small birds that travel in flocks during the day in autumn and winter roost separately, and they return to their territories where they have roost sites that they use regularly, including old nests of larger species. Coal Tits may make their way back to pine woods where some roost among bunches of pine needles.

Long-tailed Tits are highly social for much of the year. Their behaviour is different from other tits as they do not use cavities and they roost together in small groups. At dusk they head for a dense bush and perch and, clumped together, sit facing inwards with tails protruding. This fluffy ball of birds helps

Treecreeper roosts in Wellingtonia bark can often be found 2–4 metres above ground on the sheltered side of the tree.

Invited to a sleepover

Finches often feed in flocks, and in autumn and winter many roost communally. This is not a secretive affair as lots come together at a suitable site.

Different species behave in slightly different ways, often appearing nervous and flighty. Some sit on the top branches for a time and call noisily. Some groups take a short flight and then return to the bushes and trees before dropping into dense cover. It is fairly obvious that this behaviour helps attract more birds to the roost. But it is a dangerous strategy as it can also attract Sparrowhawks.

Long-tailed Tits will huddle together while roosting. This group of young huddle while waiting to be fed.

conserve energy as they keep each other warm. Roost sites are often in use for several weeks.

WHO GET THE BEST ROOSTS?

It has been noticed that the survival of female Great Tits is not as high as males (annual mortality for males is 44% and 52% for females). One factor may be the competition for roost sites in winter. Great Tits prefer enclosed sites such as natural holes and cavities, and those that cannot find one will roost in more open sites such as dense shrubs, where they are more vulnerable to cold and predators. It is likely that the larger, stronger and more aggressive males usually get and defend the best roosts.

COMMUNAL ROOSTS

The benefit of communal roosts appears to be security – birds feel safer in the company of others and their chances of being caught are less if they are surrounded by other watchful eyes. Also young, less experienced individuals and migratory birds arriving from other localities benefit from the experience of others in finding a safe and sheltered place to roost.

Perhaps one of the most important functions of a communal roost is as an 'information centre', enabling birds to head out in the morning with others who have the experience of where to feed. However, how this works in large roosts such as Starling gatherings is far from clear.

For species that roost communally, the position in the roost is important. As we have seen, the oldest and strongest will seek out the best perches, usually in the centre of the roost. Rooks also squabble over their positions and, again, the oldest and strongest tend to win.

The favoured positions are not only in the centre of the Rook roost, but also high up. One might think that perching high in tall trees would be a disadvantage, until one imagines the problems of droppings from the birds in the top tier! Soiled plumage of birds lower down is less efficient for insulation and matted feathers may even restrict flight.

Outside the breeding season Marsh Harriers sometimes roost communally. Large numbers favour some sites which they approach as dusk falls.

Some of the most dramatic roosts for birdwatchers are those of raptors. Large birds of prey sometimes roost communally, especially in winter. Harriers are especially well known for their winter roosts. A few years ago, on a cold January evening, I watched as 71 Marsh Harriers entered a communal roost in a reedbed on the Norfolk Broads. Hen Harriers roost communally on moors or heaths, but their gatherings are generally of fewer than 20 birds, although larger roosts have been recorded.

The harriers arrive at dusk, approaching low over reeds, heather or other vegetation and then drop into dense cover where they will stay until morning. Sometimes other species of raptor join them.

Wader roosts frequently comprise more than one species. Their estuarine feeding areas are used by many different species, but all are forced off by the high tides and onto safe and sheltered places above the high tide line, and sometimes onto surrounding farmland. Here different species will mix together, although generally the species tend to be in separate zones within the roost. Knots will roost next to Oystercatchers and godwits will stride among the smaller waders such as Dunlins or Redshanks. All will return to the mud as soon as their feeding grounds are revealed again.

As we saw on page 48, Pied Wagtails are territorial in summer and may have individual feeding territories in winter. They also form large winter roosts. The birds visibly gather near suitable roost sites before dusk, with considerable calling. Surprisingly many of these sites are manmade or in towns.

For years a large Pied Wagtail roost gathered in Hammersmith Broadway in West London. They also use sewage treatment works and other buildings: I once watched a very active roost at the Scratchwood Services on the M1, just north of London. Here, the birds entered a courtyard that was enclosed by glass on four sides and lit with bright lights.

Different species of waders may roost close together when driven off their feeding grounds by high tides.

Pied Wagtails roost communally outside the breeding season. Some roosts contain hundreds of birds and they may be in towns where the temperature is higher than the surrounding countryside.

It would seem that as well as security these wagtail roosts benefit from higher temperatures than surrounding areas of open countryside. Perhaps more importantly, as we saw with Starlings, these roosts may be 'information centres' and help individual wagtails, often incomers to the area, find food and shelter more quickly.

A study of a very large wagtail roost, up to 4,700 birds, in southeast England indicated population control might be another function for these roosts. There sometimes comes a point with large roosts that there are too many birds for the amount of food locally, and the mass flights and calling birds at roosting times helps birds sense the size of the population. This may encourage some to move on to new feeding areas.

A particularly incongruous roost is sometimes used by Cormorants, which have begun roosting on overhead cables and pylons. Some of these are inland, but close to lakes and rivers and other feeding areas and pylons.

Birds and weather

'**Nice weather for ducks**' is a popular idiom to describe
a downpour. Swallows seen flying close to the ground is
enshrined in folklore as warning of rain on the way, and gulls
seen inland were once thought to forecast bad weather. Until
modern times the Mistle Thrush was usually called the 'storm
cock', and there were several variations of the name 'rain bird'
given locally to the Green Woodpecker, both in Britain and on
the Continent. Several species in North America, including
Slate-coloured Junco and White-crowned Sparrow, have the
same local nickname.

WEATHER FACTORS

The weather link between birds and humans continues today,
as many people show their concern by feeding birds in winter,
and especially when the weather is particularly harsh. Such
actions can help local bird populations to survive.

Swallows often fly low
before rain, as atmospheric
conditions encourage hatches
of insects which are most
numerous low down.

The Song Thrush, expert snail-cracker, has this option if dry weather makes worms hard to find.

Weather is, however, a much larger topic than how birds survive a spell of snow or frost. All through the year, weather conditions control the food available and, for many species, nest-building has to take account of protecting eggs and chicks from the effects of heat, cold, wind and rain.

Heating up

Weather conditions help birds fly, as we saw on page 150. Warm conditions set off upcurrents, known as thermals, which save energy for large birds as they soar on outstretched wings, gaining height with little effort. Insects are also carried aloft on these currents and become food for aerial insect-eaters such as Swifts.

Sunny weather in late winter and early spring encourages some birds to display – especially the large raptors. Golden Eagles prefer bright blustery days, often following a depression, for their soaring flights over their territories.

A series of mild winters is obviously beneficial, especially to small birds, but it can lead to greater competition for food as the populations grow. Through the 1990s and into the next millennium, Long-tailed Tits survived winters rather well, and their population increased significantly. During this time many more began to visit British gardens, which may be an indication of more suitable food on offer, but also reflected their need to expand into new feeding areas.

Coastal nests on bare rock faces are exposed to both the chill winds from the sea and the heat of the summer sun. Shags are often seen 'panting' in hot weather.

Warm weather in summer is usually beneficial, but for birds like the Blackbird it can make feeding young more difficult and limit the survival rate if worms are difficult to find in hard soil.

Birds do not sweat, but in hot weather they can sometimes be seen panting, typically with bill open, which allows respiratory evaporation. Their feathers are often ruffled, and they back the wind so the cooler air currents can reach beneath the feathers.

Some species, such as Gannet and Cormorant, have an extendible pouch of skin below their bills and this is often fluttered in hot weather. Birds of prey, such as Sparrowhawks, will sometimes shield their young from direct sun by partly spreading their wings over them, in a position often called 'mantling' – the same posture is also used to guard a kill.

Some birds actively seek out sunny and sheltered spots for sunbathing. This may be pleasurable for them, but probably has a more important function as we shall see later.

Cold weather

Cold wet spells in summer are disastrous for small birds if this coincides with the time their young are due to leave the nest. A few cold, wet days in early June will wipe out many fledgling Blue and Great Tits, and we see the resulting decline in our gardens the following winter.

Invertebrates, food for so many waders and other species, become less active as temperatures fall in winter, but those on the coast are warmed by the daily tides and are more active than those inland – even in midwinter. Birds such as Curlew and Dunlin, moving from upland hills to sheltered estuaries, can take advantage of this supply of food.

When burrowing invertebrates dig themselves deeper in cold weather, the length of a bird's bill is critical as to whether it can reach them or not. Birds with long bills are more successful than ones with short bills. Bill lengths vary, even within the same species, and female Oystercatchers, which usually have longer bills than males, can sometimes continue to feed while the males are forced to move elsewhere.

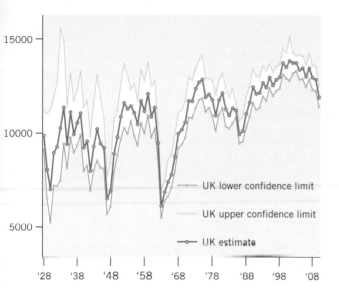

Heron highs

The Grey Heron population has been monitored regularly since 1928 in Britain. The dips in the population graph clearly show the dramatic effect of cold winters. The increase in recent years not only reflects mostly milder winters, but also cleaner inland waters, better legal protection and new food supplies around gravel workings and fisheries.

Winter chill · A long period of cold and frosty weather in winter makes food more difficult to find for many species. If harsh conditions persist, Lapwings and Golden Plovers will make long-distance 'cold weather movements', usually in a southwesterly direction, in the hope of finding more clement weather. Other species, such as Redshank, mostly stay in their wintering areas with the resulting high mortality.

In 1937, many Fieldfares made a cold weather movement out of Norway and were taken on the strong winds west towards Iceland and Greenland, where many were found dead. However, 10 years later a small colony of breeding Fieldfares was found in southern Greenland, presumably the result of the earlier cold weather movement.

Hard winters can drastically reduce small bird populations. The Dartford Warbler declined by 98%, to only 12 pairs, after the extremely cold winters of 1962–1963 and 1963–1964 in Britain.

Redwings breed near and within the Arctic Circle and are winter migrants to much of western Europe. If they encounter extreme weather conditions in winter they make cold weather movements to attempt to find better feeding areas – usually towards the south or southwest. In the unusually cold winter of 1963 one Redwing was found on a ship in the Atlantic, 1,000km northwest of the Azores.

For some species, territories become less important in cold weather, as the need to forage becomes paramount. In difficult feeding conditions even Robins, notoriously aggressive to their neighbours, will tolerate other feeding Robins close by.

Wrens are territorial for most of the year, but in cold conditions they will roost together for warmth – with a higher 'body mass' a greater amount of heat is conserved by huddling together. On winter evenings when temperatures are falling they move towards a communal roost, where they will spend the night, often in a surprisingly small space.

To survive cold weather, birds need additional deposits of fat. There is evidence this can be added quickly as temperatures fall, and birds in cold weather generally weigh more than in mild conditions. Those that do not manage this weight gain are probably the ones to succumb.

Snow · A blanket of snow hides food, although sometimes areas of bare soil remain around shrubs and in woodlands, where birds continue to feed. Some species are equipped to deal with snow, and the Ptarmigan, which lives year-round at high altitudes, digs into the snow to find its food, and will roost under drifts of snow for warmth.

Rather than move off the hills after a heavy fall of snow, Ptarmigans may go even higher! This rather contradictory

On some lakes, the constant movement of many wildfowl keeps the water moving and helps to prevent more ice from forming.

Pyjama party

Wrens roost communally in cold weather and conserve heat and energy by huddling together. An amazing record of 96 Wrens were seen entering and leaving a small hole leading to a loft space in a house situated close to English woodland during cold weather in 1979. They started to arrive 25–45 minutes before darkness, and on occasion they arrived in such numbers they had to line up to get in!

Another report was of 63 Wrens entering a single nestbox, again during cold weather. The owner was so concerned that another nestbox was erected and 12 more Wrens went in that as well.

movement is explained when considering the effect of wind on mountains, which blows snow from the mountain tops and builds up drifts in the valleys. The Ptarmigan can withstand the cold as long as it can reach its food of leaves and shoots.

Ice · If normally ice-free lakes freeze over, the wildfowl that depend on them are driven to find open water elsewhere. Those on smaller lakes, which freeze first, move on to larger lakes and reservoirs where freezing is less likely. If freezing conditions continue many will be forced to leave the locality completely and visit estuaries and sheltered coastal waters.

If ice forms on the inland waters where they live, Dippers and Kingfishers will move downstream to larger rivers and sometimes to estuaries.

Sometimes rain freezes and covers branches and the ground. This 'glazed ice' makes finding food impossible for small birds, and if it persists will cause high mortality.

Estuaries have low salinity and less wave action than the sea, and often have sheets of ice on them after nights of low temperatures. Redshanks are badly hit at these times, when ice covers their feeding areas, and their food of invertebrates and

Two steps forward, one step back

Cetti's Warbler, which began colonising Britain in 1972, was almost wiped out again by several cold winters in southeast England between 1978 and 1988. It recovered and continued its expansion until the winter of 2010–11.

crustaceans is either covered with ice or goes too deep in the mud for them to reach.

Wind

One obvious effect of wind on birds is the way they arrange themselves in open areas. Watch a flock of Lapwings or gulls standing in a field on a windy day and they will all be facing into the wind – this way the wind does not ruffle their sleeked down feathers. Any other direction is not only unpleasant, it is likely to result in their bodies cooling more quickly as the wind gets under the feather coating.

In uplands, winds and air currents carry small insects from lower areas and deposit them high on the hills and even on snowfields – providing food for upland species such as Snow Bunting and Meadow Pipit. Wheatears and pipits often move further uphill to take advantage of this food supply.

Strong wind will change the behaviour of small birds. Tits will not feed as high in trees, and Bearded Tits will stay low within their reedbeds, whereas in calm weather they can often be seen flying over the tops of the fluffy reed heads.

Many waders feeding in exposed places, such as estuaries in winter, experience wind chill. They also lose feeding time due to high onshore winds covering feeding areas under high tides.

Gales are another problem. Much depends on the time of year, and obviously the nests of treetop nesters, such as Rooks, Carrion Crows, Grey Herons and Ospreys, are vulnerable to damage or even destruction.

Ravens' nests on exposed cliffs have been known to be blown off in extreme conditions. It says much for the construction of the nests that most withstand gale-force winds, and survive intact from year to year. Losses of young due to wind damage are limited.

Storm-force and hurricane-force winds are unusual in Britain and western Europe. But the well documented storm-force winds of 1987, which swept parts of southeastern England, left a trail of uprooted trees and destroyed woodland. As it happened in autumn, breeding birds were not affected, but much habitat was destroyed – at least in the short term. Conversely, Nightingales, Nightjars and some other woodland species actually did better in the following years, due to the newly created clearings.

In North America, hurricanes and tornados are more

Storm damage temporarily changes habitats. Dead wood provides homes for invertebrates and additional light changes ground vegetation. Damaged trees provide opportunities for new nest-sites as wood starts to rot.

Fragile foundations

In 2004, storms and rain almost prevented one of the first pairs of Ospreys from trying to nest in Wales in modern times. The pair were young, and therefore inexperienced birds that were not building on the foundation of an existing Osprey nest. In the first year storms and high winds caused their nest to capsize after the eggs had hatched with the result that the young tumbled out and died.

The following year the birds returned to find a strong artificial nest constructed by the RSPB. This was used as a base for their new nest and the young Ospreys were successfully reared.

Storm damage

Scientists studying a very stable Crested Tit population in a forest in Belgium noted that after a severe storm, which uprooted 25% of the trees, the population dropped by a half. They also showed the new population had many new birds within it and even old birds that re-nested had different partners.

frequent. There was one reported incident in 1993 in Louisiana when an estimated 40,000 birds of 45 species were killed after a night of migration. A massive 200,000 dead birds were washed up on the shores of Lake Huron in 1973.

Storms at sea · A great many birds spend most of their lives out at sea. Arctic Terns move from northern to southern hemispheres, but most terns and skuas move to tropical or sub-tropical waters, and others like the Fulmar and Gannet remain in northern waters all winter.

Many seabirds can continue to fish and feed, even in rough conditions. Indeed, the disturbed water may help, as the birds are less likely to be seen by their prey. Storms at sea are visible from some way off, and birds may be able to take avoiding routes. Some birds simply 'sit out' a storm by swimming and resting on the surface and thus conserving their energy.

Away from shallow coastal waters, the sea conditions below the surface are not much influenced by the weather, and those species that dive and hunt under water like the Razorbill and Guillemot can continue to feed.

Grey Phalaropes are waders, but after breeding they leave land and spend winter on the sea. They are normally only seen in western Europe after westerly gales have driven them ashore.

Above the surface the chief element is the wind, and seabirds can use this to save them energy by effortless gliding flight or even flying into the wind and hovering when hunting – a technique used by the terns.

The tiny petrels can fly in the troughs between the waves where the airflow is minimal, even in rough weather. From time to time, though, they get caught by sudden shifts of winds and become displaced – once above the waves they may become disorientated. This results in 'wrecks' of oceanic seabirds ending up in inshore waters. Species like Leach's Storm-petrel and Sabine's Gull may then be seen in harbours, estuaries and even on inland lakes near the coast – especially in western Britain.

Storms at sea result in more coastline being covered for longer – meaning less feeding time for birds such as Purple Sandpiper and Turnstones

A time to reap and a time to sow

Weather affects the time of sowing and harvest, and this influences the amount of food birds can obtain from agricultural fields. Late harvests and more fallen seed results in more food for seed eaters such as Yellowhammer. Late ploughing can reveal invertebrates that are food for gulls and Lapwings, which can frequently be seen following a tractor and plough across open fields.

Land gulls

Gulls are more common inland in autumn and winter and they benefit from autumn ploughing, when farm machinery cuts into the soil to reveal a supply of invertebrates. Weather affects the time of ploughing and the amount of food available to the birds.

Drought

One of the most dramatic hazards in a period of drought is the potential for fire on heaths and moors. While occasional autumn burning can be beneficial, fierce burns in summer not only kill young birds (and many other creatures) but often do long-term damage to the underlying peat, so that the heather does not return.

In Europe, thrushes and other species looking for worms and other invertebrates in the soil are hard pressed at this time and may raise fewer young, but on the whole the effects on populations do not seem to have been too serious. For migrants, though, drought is much more serious.

For years the Sahara desert has been expanding due to drought and human influence. This means that birds migrating to sub-Saharan Africa for the winter have even more inhospitable terrain to cross. We are now seeing fewer African migrants returning to Britain and Europe each spring. We do not understand the reasons, but drought conditions in North Africa is likely to be a contributory factor.

Rain

Rain causes mud, which is used in the construction of bird's nests. Dry spells in spring can be a problem for Swallows and House Martins in particular. Some species use wet nest material, and Wrens depend on wet material as this shrinks to help form their compact nests of mainly moss.

Rain restricts hunting by large birds of prey, and Buzzards and eagles will sit tight on days of heavy rain. However, in front of rain showers and thunderstorms are atmospheric conditions that assist other species – especially those that feed in flight. Radar has shown Swifts feeding at high altitudes in association with thunderstorms, where their aerial prey has been lifted by the atmospheric conditions. Swifts may follow their prey up to a height of 900m and, depending on the feeding and weather conditions, they may travel 800km from their nests.

In Africa, Hobbies feed on flying termites which emerge after thunderstorms, and there is evidence this falcon deliberately moves from storm to storm, and can even detect storms from as far away as 160km.

Adult plumage of most birds is resistant to rain, and ducks have especially good waterproofing. Ducklings, however, are downy and vulnerable to chilling, and will sometimes shelter from rain. Wet ducklings that are inactive and thus not generating heat can quickly lose their body warmth and may become chilled and die.

Rain is also a hazard to migrating birds. If caught in rainfall over the sea they face possible death, as shown by hundreds of Garden Warblers washed up on a beach in Spain in 1991 after a night of heavy rain.

Some species, such as Woodpigeons, are often seen bathing during showers of rain as we will see on page 227.

Flooding · After heavy rain, flooding can make a serious impact. Birds that nest on islands in lakes, such as Common Tern and Little Ringed Plover, may find their nests are swamped by rising water levels. Godwits on flat marshlands

Swifts have been known to move 1,500km from their breeding colonies ahead of thunderstorms.

need wet meadows for feeding, but too much water can ruin their nest-sites. However, wildfowl such as Teal and Garganey, nesting in similar places, may benefit from spring flooding that brings increased feeding areas for their families. Black-necked Grebes may colonise newly and temporarily flooded areas.

Great Crested Grebes have nests that are attached to vegetation growing in the water, but their nests have some ability to float with rising water level. Coot nests are more likely to be swamped – especially by a combination of wind and wave action.

In addition to normal nests Moorhens often construct 'brood nests' for their newly hatched and fluffy young. One remarkable observation was of a brood nest that was built quickly and the young transferred to it when the original nest was in danger of flooding. The young returned to their original nest once the water level fell again.

WEATHER FOR MIGRATION

In general, birds on migration will fly with the wind behind them. In western Europe, as low pressure moves to the east, or high pressure arrives from the west, northerly winds are created which will assist migrants heading southwards in

Black-tailed Godwits breed on flat marshland, but their nesting can be ruined by occasional summer flooding.

autumn. In spring, migrants will mostly arrive with the help of warm southerly winds caused by high pressure moving to the east or low pressure approaching from the west.

Nights of good visibility, bright stars, no cloud and a following wind are the best conditions for long-distance migrants to set off. If they meet cloud they are likely to become disorientated, but if they can get above the cloud they will see sun or stars, and below it they may have visible clues from the ground. Many will land if forced down by rain or low cloud – if the weather closes in over the sea, they may turn around and head back to land.

A tailwind will save a bird energy as it will fly slower. It will lose more energy with a headwind as it will fly faster. If the wind is faster than the bird's flight, then it is in danger of losing control or overshooting its target.

Some migrants will be displaced by crosswinds. Many, especially older birds, will be able to re-orientate themselves. Some will get carried far off course to become vagrants in unusual areas, to the excitement of those birdwatchers known as twitchers, who particularly seek out rarities.

Temperature is only important in that it may influence food availability, and higher temperatures may help birds to increase their body fat to the levels required for their journey.

A sudden temperature drop after migrants have returned to their breeding grounds can be disastrous. Lapwings returning to breed in Norway were reduced by 30–80% in different regions as a result of unseasonal snow and frost in April, as many died or abandoned their breeding attempts.

Extreme weather events will potentially kill a great many migrants. Poor visibility in snow, fog, rain or a combination of these often spells disaster. At these times birds fly low and become disorientated, and they may collide with human structures such as radio masts, tower blocks and gas flares. If bad weather strikes over the sea, many are doomed. Some find ships or oil rigs to land on, but this must be only a tiny proportion of the total, and whether even these can take off, re-orientate and survive after this, no one really knows.

An estimated 100,000 King Eiders died on their return migration in Canada in the spring of 1964, when the sea re-froze after having started to thaw.

Plumage and moult

Juvenile Starlings are uniformly dull brown, but in late summer they begin to moult into their dark and spotty adult plumage, starting with their undersides.

Birds are the only living creatures to have feathers. Intricate, strong, yet remarkably light and flexible, a single feather is a thing of great beauty. Feathers are essential for birds' adaptation for flight; they provide insulation from cold and their colours and shapes create a canvas on which can be displayed an almost endless variety of patterns and hues.

BEAUTIFUL PLUMAGE

Plumage is the name given to the complete 'coat' of feathers that covers a bird's body. The number of feathers varies between species, but a hummingbird has fewer than 1,000 individual feathers, and a swan more than 25,000.

Plumage patterns and colours help birds to recognise their own species, allow them to tell male from female and adult from juvenile and, of course, it gives us a chance to identify them, but there is much more to it than that!

THE VALUE OF FEATHERS

Wing and tail feathers provide power and control for flight. The feathers covering the body give protection and reduce friction,

Deadly decoration

It was once fashionable for women to wear feathers, especially in hats. Tons of feathers were imported for the fashion industry in the late 19th century, many from egrets – shot during the breeding season with their young left to die. Campaigns against this cruelty led to the formation of the RSPB in Britain and also the National Audubon Society in the USA. The RSPB began life as 'The Plumage League' in 1889, founded by a group of very determined women in Manchester. In the USA, violent confrontations occurred over bird protection, and in 1905 a ranger was shot while guarding egrets in the Everglades.

All the frills

Grebe display starts in late winter, and both male and female grow the ornate head feathers that are used in ritualised display. The breast and belly feathers are incredibly dense and were once used as 'grebe fur' for making muffs in Victorian times.

enabling high-speed flight, or (in diving birds) fast underwater swimming. The dense and oily feathers of ducks, combined with a thick layer of down, contain air that increases buoyancy and insulation. The soft plumage of owls allows them to fly silently and increases their hunting efficiency.

As much as 95% of a bird's plumage may be made up of captured air. The combination of air and feather helps to regulate temperature and keep a bird's body dry, as well as aiding flight.

The insulation qualities of feathers have been exploited by humans. We keep warm under soft and light duck-down quilts. Duck skins were incorporated into clothing by early Inuit communities, and the harvesting of 'Eider down' for bed coverings continues today. This is not as cruel as it might sound, as female Eiders pluck their own breast feathers for their nests, and it is this material that is collected after the breeding season. The insulation properties of the Eider, which breeds on the ground in the Arctic, is higher than that of other ducks and geese and remains a very valuable commodity from a sustainable source.

Communication with feathers

The colour, shape and pattern of feathers can be used to send signals to others. Robins display their red breasts as a warning to intruders. In spring Great Crested Grebes expand their chestnut head frills and tippets as pairs display to each other on lakes and rivers. Male Mute Swans arch their flight feathers over their backs, giving them the impression of intimidating extra bulk to help drive off an intruder, and Lapwings tumbling over their territories make a noise known as wing-strumming, caused by the vibration of their outer primary feathers as they plummet earthwards.

Perhaps even better known is the male Peacock, spreading his magnificent feathers in a grand display to impress females. In fact these are not true tail feathers, but are the tail coverts. In most other species, the tail coverts are very insignificant

The cryptic camouflage of the Woodcock's plumage makes it hard for predators to find when nesting on the woodland floor.

contour feathers forming a small pad on the upper tail base, but in the Peacock they are elaborate, bearing large and colourful 'eye' markings, and are greatly elongated.

Feathers for hiding

Plumage can very successfully disguise a bird. The delicately marked feathers of a Woodcock help it to merge with the ferns and dead leaves of the woodland floor. In contrast, some birds which live in open country, such as the Ringed Plover, have plumage that is boldly marked and helps to break up the shape of the whole bird. This 'disruptive camouflage' turns the bird into a random shape in the landscape.

Many birds have a dark upperside and lighter underside, a pattern known as counter-shading. As the upperside is naturally in light and the underside in shadow, counter-shading evens out the overall tone to make the bird less noticeable as a solid object. Some waders may even reflect the ripples of moving water on their white breast as they wade along.

Sexual differences

With some species it can be virtually impossible for us to tell the sexes apart. Male and female Robins look alike, and even the birds find it difficult to tell – it is their behaviour that generally gives the game away. With other species, male and female have completely different plumage, and in some cases have even wrongly been classified as different species, before careful behavioural observation revealed the truth.

In many cases where the sexes are different (sexual dimorphism), the males are highly coloured to attract a mate but take no role in looking after the resultant eggs and chicks. The females are dull-coloured and cryptically camouflaged to protect them from predators while sitting on a clutch of eggs. In a few species, the roles are reversed. It is the male Dotterel and the male Red-necked Phalarope that are duller and incubate the eggs, while the brighter females lead in display and have little to do with rearing the family.

Hoopoes raise and lower their crests during display, when driving off rivals or when alarmed. Sometimes it is raised briefly as a bird lands.

A young Herring Gull will not completely resemble an adult until it reaches its fourth year.

Age differences

Some birds hatch with a covering of downy feathers while others are naked. The first full plumage is often different from that of the adult, and is usually duller and more cryptic. Juvenile Blue Tits and Great Tits are like washed-out versions of their parents until their first winter. Generally, the body feathers will be moulted into adult plumage before winter, but flight feathers are retained until the next season.

Juvenile plumages often provide added protection. The juvenile Robin is speckled, giving additional camouflage in dappled sunlight, and it lacks the red breast – preventing it from being attacked as a rival by other Robins. In sexually dimorphic species, juveniles often closely resemble adult females. Winter gatherings of sawbill ducks like Goosanders and Smews usually include more 'redhead' female-like birds than striking adult males, but the redheads will be a mixture of adult females, and young birds of both sexes.

Some species take several seasons to grow their adult plumage. Most juvenile gulls have mottled grey-brown feathers and take several years to mature – becoming whiter with each annual moult. Gannets have several distinct immature plumages and may retain traces of juvenile plumage until their fifth year.

Seasonal differences

Some species take on a different look, depending on the season. Generally this is referred to as summer and winter plumage, but for many, 'breeding' and 'non-breeding' plumages would be a more apt description. Black-headed Gulls lose their dark brown head feathers between July and September and have mostly white heads until about February, when they begin to grow their brown feathers ready for the new breeding season.

The Ptarmigan, a grouse which lives in mountainous regions, has a more complicated seasonal plumage sequence. In summer, its beautifully patterned grey, brown, black and white plumage blends with the lichen-covered rocks. In winter

Northern trend

Some Guillemots have white spectacle-like markings and are known as 'bridled' Guillemots, although they are the same species and interbreed with normal Guillemots. While relatively few bridled Guillemots are found in Scotland, the proportion of 'bridled' birds increases in Iceland and northern Scandinavia – from 5% in the south of its range to 50% in the north.

it becomes pure white to match its snowy habitat. Spring and autumn plumage is intermediate grey and white.

Polymorphism

When a bird species consistently shows two or more distinct colour forms within a population, this is known as polymorphism. This is not common in birds, but there are some notable examples. Arctic Skuas may have light, dark or intermediate plumages (often called 'phases' or 'morphs'). The different morphs regularly interbreed, but there is a tendency for the lighter ones to nest further north, perhaps gaining benefit from the lighter plumage, helping their camouflage among late-thawing snow and ice.

A flock of feathers

Most feathers have a shaft and vane, and their shape and size varies greatly.

Flight feathers • The longest and strongest are the primaries and secondaries in the wings, and the tail feathers, which allow a bird to create the dynamic forces for flight.

Contour feathers • These are the smaller feathers that cover the bird's body and wings. Many are brightly coloured towards the tip, and together form much of the bird's pattern and colour.

Down and semiplumes • These tend to be small and fluffy, with many of the barbs lacking barbules and so not able to interlock. These are the feathers closest to the body and hold much of the air that will help keep the bird warm.

Bristles • Around the eyes and mouths of insect-catching birds such as flycatchers and Nightjars are small feathers that have evolved to resemble short thick hairs. These are known as rictal bristles. It was assumed these helped the bird catch or trap its prey, although it now seems more likely that they help protect the bird's eyes.

WHAT IS A FEATHER?

Feathers are made of a strong protein called keratin and evolved from the scales that covered birds' reptilian ancestors. Like our own fingernails, full-grown feathers are dead tissue, and they are very light and amazingly strong.

Feathers have a hollow central shaft with vanes on either side. These vanes are made up of separate branches known as 'barbs', from which tiny 'barbules' branch off and interlock with minute hooks and ridges. It is this structure that allows a feather to be 'unzipped' and repaired like avian 'Velcro'.

Feathers can be long like those that make up a Pheasant's tail, or short like a bristle round a Swallow's bill. Some small down feathers regularly break up and produce waxy, scale-like particles known as power-down. This can sometimes be seen when a bird collides with a window and leaves a ghost-like impression behind. Most often these 'ghosts' are made by pigeons or doves (see page 225), but sometimes by owls.

Colour ways

It is the colour of birds which many people find so eye-catchingly attractive: the rosy pink of a male Bullfinch's breast, the flash of yellow in the wing of a Goldfinch or the gaudiness of a male Pheasant — there is much beauty in nature, and birds contribute to it in so many ways – not least in their plumage.

Colour comes from either the physical structure of the feather, from chemical properties giving pigmentary colour, or a combination of the two.

Structural colour • Many colours in a bird's plumage are due to iridescence, as certain wavelengths of light are reflected from flattened or twisted barbules. Often we see different colours depending on the angle of the light on the bird. The head of a male Mallard may look blue or green depending on the light,

The blues and greens of a Kingfisher's plumage are the result of light diffracted off what are actually dull-coloured feathers. Hold a Kingfisher feather against strong light and it looks plain blackish.

The black flight feathers of White Storks contain melanin that helps give them strength and prevent wear.

and this is caused by structural colour. Swallows, Magpies and Kingfishers all have some iridescence in their plumage.

Birds may also possess non-iridescent structural colour, formed by the scattering of short waves of white light by very small particles. These colours do not appear to change depending on the angle of light. One example is the powder-blue cap of a Blue Tit.

Pigmentary colours · The three pigments that give birds colour are melanins, which produce black, brown, reddish and yellowish tones, carotenoids, which produce bright reds, yellows and oranges, and porphyrins, which result in red, green and brown colours. Melanin adds strength to feathers, and many birds that have white plumage have black wing-tips to give additional strength. Porphyrins tend to be light-sensitive and fade in strong sunlight.

Some pigmentary colours can be picked up from a bird's diet. Flamingos derive carotenoids from the algae and crustaceans they eat, and this colours their feathers pink.

Combined colours · Some colours are the result of the combined action of two pigments; Greenfinches are green because of the combination of black melanin and yellow carotenoid. Combinations of pigment and structural colour create even more colour variety.

Unusual plumages · Visit a museum that displays stuffed birds in glass cases and you are likely to see some familiar birds with 'wrongly' coloured plumage. Old collectors and curators relished these anomalies or 'sports' for their rarity – like a stamp with the queen's head upside down! Such specimens fascinated me as a child, when visiting the hallowed halls of the Natural History Museum in London.

All-white animals with pink eyes are albinos, lacking all melanin. True albinism, the result of a genetic mutation, is rare in wild birds as there are usually severe eyesight problems associated with the lack of melanin in the eyes. In other animal groups where eyesight is less important to survival, albinism is observed more often. Carotenoids are unaffected by albinism, so an albino Goldfinch, for example, still has its red face and yellow wing-bars.

White Puffins are mentioned in folk legends and one is reputed to have lived for 50 years. White Puffins are either albinos or, more frequently, as with this bird, leucistic.

White birds with normal eye pigments are more common, as are birds with white patches in otherwise normal plumage. Such birds are generally known as leucistic, and survive better than albinos as they have normal vision – though lack of camouflage makes them vulnerable to predation. Other pigment abnormalities may include the absence of brown but not black melanins, producing a grey-toned bird, or dilution of all pigments, giving a 'washed-out' appearance.

CHANGING FEATHERS

Feathers may be strong, but over time they become abraded and less efficient. Most are replaced annually, generally after the breeding season. The process of shedding old feathers and growing new ones is known as moult.

Losing too many feathers at once would compromise a bird's ability to fly and keep warm. Therefore, birds lose their body feathers only as new ones grow, and flight feathers are replaced

A Cormorant in wing moult.

in a sequence, usually from the centre of the wing outwards. This ensures they do not lose the power of flight. However many water birds and some seabirds do have a short period when they are flightless.

Different strategies for moulting

Birds have evolved different times and patterns for moult that suit their annual cycle of activities – resident and migratory species have different needs, as do large raptors compared with small songbirds.

The basic pattern is for birds to moult after the breeding season. By then their plumage has become worn by the efforts of rearing their young, but once the chicks are independent then the parents can undergo the new rigours of moulting. Growing new feathers requires extra energy and food must be plentiful. Song ceases, or is greatly reduced during this time. Some migrants complete their moult before flying south. However, some will delay their moult until they reach their wintering grounds.

Some birds suspend their moult for a time. Having started moulting they then stop, make some or all of their migratory flight, and then recommence their moult in a food-rich area. Wading birds such as Grey Plover and Knot do this on British and European estuaries; they are migrants from the Arctic and many use the rich food source of the estuary as a feeding and moulting area before moving on down the coast of Europe to Africa.

Moult may begin while some species are rearing their young, perhaps because there is a ready supply of food, or the breeding period changes for some reason. Snow Buntings nesting in the Arctic have to start to moult while feeding young in order to have a complete new plumage before the harsh Arctic winter weather sets in.

Prenuptials

For birds that have a distinct breeding plumage, like the male Ruff or the Black-headed Gull, there is usually a second partial moult of head and some body feathers.

As egg-laying approaches, more feathers are lost from the centre of the breasts of birds that will incubate their eggs. This exposes the brood patch that delivers the necessary warmth to the developing embryos, and the shed feathers may be used to form part of the nest's soft lining.

Avoiding a second moult

Stonechats, Reed Buntings and many other small birds moult in late summer into a dull streaky brown plumage for winter. In spring their colours are brighter and the males have jet black head feathers, yet this second transformation is made without expending the energy to moult. The new autumn feathers have pale tips which overlay the brighter plumage underneath. As spring approaches the pale tips abrade away and the bright spring plumage is revealed.

Total eclipse

In summer, male Mallards (and males of many other duck species) moult from their bright breeding plumage into a much duller, female-like plumage. This plumage is known as 'eclipse'. They also moult all their flight feathers simultaneously, and

Spot the difference

Starlings moult into their spotty winter plumage between June and October, but as spring approaches, and the birds start visiting potential nest-sites, the pale spots which lack melanin and are weaker start to abrade away, revealing the brighter iridescent spring and summer plumage.

so become flightless. The drab plumage gives them a cryptic camouflage to protect them during this flightless period. Before winter they have a further moult back into bright breeding plumage, and by October and November they are looking their best ready for courtship.

Drake Mallards in their eclipse plumage. During this annual moult their flight feathers are moulted in rapid succession and they are flightless for about four weeks.

Staying airborne

Larger birds, especially the bigger raptors, have protracted periods of moult. Golden Eagles, for example, moult between April and October and some may continue into winter. A complete moult may take two seasons, and some feathers are retained for up to three years. It seems that primary feathers are replaced more frequently than secondaries, presumably because these are the feathers that have the most wear.

Feather care

As we have seen, feathers are unique to birds. They are essential to their lives, not only for flight, but also for insulation and communication. It should not be a surprise that birds spend a lot of time looking after their feathers, and that the care of their plumage has resulted in a number of distinct behaviours, the most familiar being preening.

PREENING

Most of us will have, perhaps as children, 'unzipped' a feather and then 'zipped' it up again. This fascinating property of feathers means they are self-repairing, and allows a bird to pass through thick and thorny shrubs without 'tearing' its plumage. There are several actions made by birds that we generally call preening, all of which help get disturbed feathers back into shape.

The most familiar form of preening is when a bird runs individual feathers through its bill. This helps straighten them and ensure the barbs and barbules interlock and form a perfect

Many birds regularly bathe, but wet plumage can limit flight and make a bird vulnerable to attack by a predator.

feather again; this is particularly important for wing feathers. Loose matter is removed, and this action also helps spread a natural waterproofing oil that is secreted from the preen gland.

Some preening involves 'nibbling' at a feather or group of feathers and helps to remove dirt and spread oil. Mallards can often be seen nibbling at their breast feathers. Some birds, such as the grebes, will wet their bills before preening.

Another action is more akin to stroking the feathers with the bird's bill, usually in a downward direction that helps smooth feathers back in place.

Preening takes place at any time of day and sometimes in the evening roost. It can be a quick rearrangement of a few feathers, or a long period of meticulous grooming. Preening also takes place after bathing and other activities associated with feather care, and can often be seen when birds are 'off duty', resting or loafing.

Sometimes one bird will preen another in an activity called 'allopreening'. Generally this involves birds that are already paired, and the allopreening may be part of building the bond associated with courtship. A male Guillemot will often preen the neck of its mate, but sometimes an individual will preen a neighbouring bird as it incubates its egg. Pigeons and doves are frequently seen allopreening, and also parakeets.

Little Owls may engage in allopreening throughout the year; this behaviour is mutual – male preening female and female preening male – and not only the neck is preened, but also the breast and even the feet. A Little Owl mutual preening session can last for up to half an hour!

Oiling

Most birds have a gland at the base of their tail that secretes a natural oil, which helps to lubricate the feathers and increase their waterproofing. Often at the start of preening a bird squeezes the preen gland or rubs its bill over it and transfers the oil to its feathers. Sometimes the oil is rubbed off the bill with the foot and then applied to the feathers.

Allopreening is the name given to mutual preening, when one bird preens another. This appears to be a social activity often linked to courtship, but must also assist in care of those feathers that birds find it hard to reach in normal preening.

Cormorants and Shags can often be seen 'hanging their wings to dry', but there may be other advantages such as thermoregulation.

DRYING

Most birds have stereotyped actions to get rid of surplus water from bathing, swimming or even diving. An Osprey that has just dived into a lake for a fish will usually make a vigorous shaking movement after rising from the water. For a moment it will lose momentum as it sheds the water from its plumage, but this only takes a few seconds and the full power of flight is soon restored.

Body shaking and wing flapping is commonly seen in water birds. They may be shedding water or simply rearranging their feathers after diving. Sometimes the purpose is obscure, and this can be best described as 'comfort behaviour', something a bird needs to do to resume normal life. Many ducks bathe, wing flap and preen vigorously following mating.

Cormorants, Shags and the darters (snakebirds) of the Americas and Africa characteristically spread their wings 'to dry' after swimming, feeding or bathing. These birds have natural oils, and while this activity may help the wings to dry it is quite possible the pose has other benefits, such as thermoregulation, or even a social benefit such as attracting other birds to the site.

A LITTLE POWDER

Some birds, particularly the heron and pigeon families, produce powder down. Pigeons have no oil gland but powder down is formed at the base of their feathers, from the cells covering the growing feather. Herons have special powder down feathers that are fragile, and break off to produce powder.

Sometimes you may see a dusty impression or 'ghost' of a bird on a window. These are formed when birds collide with the glass, and their powder down leaves a striking 'negative image'. Depending on the angle of the strike the image can be so complete that the head, breast and wings can all be recognised.

Herons tend to get very messy as they fish, especially those that catch eels. A captured eel often tries to resist being eaten by wrapping itself around the bird's bill or neck. This leaves scales and slime on the heron's feathers. To help clean itself the fouled bird will rub its head on its breast where the powder is collected and then groom its feathers with its claws.

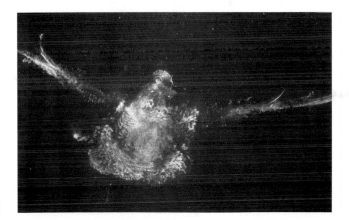

Powder down is left on a window after a bird has collided with it. This 'ghost' was made by a pigeon.

A personal comb

Herons' feet are unusual in having tiny serrations on the inner edge of the middle toe. This is known as a pectinated claw, and when used in a combination with powder down, helps dry and clean the plumage. Nightjars also have this type of claw, although in these birds purpose is not really understood.

SCRATCHING

A bird may suddenly scratch in response to a sudden irritation, but scratching is also part of feather care. Head scratching is the most usual way for a bird to preen its head feathers; this

Grey Herons take a wide variety of food, especially fish, and many take eels where they are plentiful. Eels present a challenge, as some will wind themselves around the bird's neck. Eventually they are swallowed down whole, and usually alive!

usually happens when a bird is perched, but sometimes in flight.

Some birds, including parrots and pigeons, are direct scratchers, bringing the foot straight up to the head. Indirect scratchers, which include most passerines, lower the wing first and bring the leg over the top of it.

BATHING

Bathing is an essential form of feather care for many species, and there are various ways that have evolved for birds to take a bath. Of course, a bathing bird can never be completely 'off duty', it must always be alert to potential danger – even more so as wet plumage may restrict its flight.

The indirect scratching method may be an evolutionary 'hangover' from birds' four-legged ancestors.

The Robin is a tentative bather, often bathing at the very edge of the water. The bird makes quick bathing movements as it flicks water onto its plumage with its head. A more full-on approach is common in most birds: they will stand or crouch in the water with fluffed up feathers and vigorously push water through their plumage by shuffling their wings. Some, such as pigeons, dip their heads into the water and then let it run back over their bodies or beat the water with their wings.

Some birds simply jump in and out of the water, but Swallows and other aerial species will 'flight-bathe' by flying low over water and dip their bodies without actually landing. A few species also plunge-bathe from a perch. Kingfishers will plunge-bathe several times in succession.

Sparrowhawk bathing. Even top predators take time out to bathe; if not disturbed this will be followed by preening.

Rain-bathing

A curious variation is rain-bathing. Pigeons and doves do this frequently. Initially, if caught in rain a Woodpigeon adopts a very upright stance and sleeks down its feathers, which results in the rain mostly being deflected.

On other occasions they deliberately get wet by fluffing up their feathers and allowing the rain to get underneath. Often they lay on one side and lift a wing, so the rain reaches the under-wing feathers. Following a rain-bath the bird usually flies to a different perch where it will preen.

Dusting

Some birds have a quite different method of cleaning their feathers that does not require water. This is known as dusting or 'dust-bathing', and not surprisingly is more common in species that originated in arid regions. House Sparrow, Skylark and gamebirds, especially Grey Partridge and Pheasant, are all enthusiastic dust-bathers.

There are three different actions required to dust-bathe. A suitable place needs to be found and some topsoil may have to be loosened before dusting can begin. Then the bird will make bathing actions in the dust, often forming a cup-shaped scrape in the ground. These actions toss dust into the air that settles amongst the feathers. Finally it draws the dust through the feathers and then shakes it out. This is like a 'dry shampoo' approach to feather care.

Sun-bathing

Pigeons not only rain-bathe, they also sun-bathe, and so do many other species. A bird will sit with its back or side to the sun and fluff up its rump feathers. Different species have their own methods of 'sunning'.

The purpose of sunning appears to be thermoregulation, as it may take place on cool days when the bird can benefit by absorbing heat directly, from the sun, which compensates for any metabolic loss of body temperature. This behaviour may be seen at any time, but mostly when the sun is high in the sky, around midday. It can occur in any season, but especially autumn, winter and spring, and is less common in summer. In summer thermoregulation may require a bird to take the opposite strategy and find a cool spot in which to rest.

There are two forms of sunning. The first, 'sun-basking', is where the bird positions itself in a sheltered spot and receives the full force of the sun's rays. The second, 'sun-exposure' is more elaborate; the wings and tail may be spread.

Birds that are sun-basking usually fluff up their feathers and sometimes droop their wings and sit in the full sun while resting. This is common behaviour with many species.

One of the most familiar species that can regularly be seen sun-exposing is the Blackbird, which may use a garden, a park or anywhere the bird remains undisturbed for a time. Always sunny, but not necessarily warm, the ideal site is often close to cover which provides shelter and refuge should a predator approach.

At first sight the bird looks unwell, as it lays on one side with its feathers fluffed up, the wings spread and the tail also spread and often twisted. The eyes are closed and the bill gapes open. It is impossible to tell if the bird is in agony or ecstasy! The position is on the ground and backing or sideways to the sun. If undisturbed this behaviour lasts for several minutes. When finished the bird preens vigorously.

Other species have different positions. I once watched a Robin beside a public path on a nature reserve, lying sideways to the sun and looking very sick. Visitors walking along the track were convinced it was injured. But later it got up and preened – it had been sunning.

Dunnocks sometimes spread a wing and allow the sun to reach the underside. Pigeons will adopt the same position

Sunning can take place at any time of year as long as the sun is bright. Blackbirds frequently engage in this activity, but it is more unusual to see a Grey Heron sunning.

A Jay will deliberately find an ants' nest, stand or squat on it and allow the ants to run up into its plumage.

as when they are rain-bathing and also raise a wing. Song Thrushes may face the sun and spread both wings symmetrically.

These exaggerated and stereotyped performances are a bit of a puzzle. They are more elaborate than is necessary for thermoregulation, and as they finish with a vigorous bout of preening they appear to be linked with feather care. There are theories that the warmth helps the birds spread the natural oil from their preen gland and that it disturbs any external parasites. It might also speed up the synthesis of Vitamin D, which would improve feather condition.

Many large birds also have sunning positions in which they spread their wings so the sun reaches the undersides. Some research has shown that strong sunlight can help restore flight feathers to their original shape.

Anting

One of the most fascinating bird habits is what is described as 'anting'. In Britain the species you are most likely to see engaged in anting are Starling or Jay. An individual will find a place with many ants, especially large Wood Ants, and will deliberately pick up one or more, squeeze them, and wipe the body fluids that the ants exude over their feathers. Most of the selected ants are workers that eject formic acid for their protection, and it is this the bird applies to its feathers.

The actions of anting are stereotyped, with wings and tail held in unusual positions as a little dance takes place, and this appears to be a routine a bird knows instinctively. The purpose is not feeding, although some ants may be eaten, but the use of the acid is assumed to be part of plumage care with the formic acid helping to discourage mites, ticks and other parasites that live in a bird's plumage.

Another form of 'anting' behaviour takes place where there is a swarm of ants. The bird sometimes settles down on the ants' nest, with its wings and tail spread, rather like sunning, and allows ants to crawl amongst its feathers. The ants, being defensive, squirt their formic acid at the intruder.

After anting the bird is likely to preen and bathe, indicating this habit is both normal and an important part of feather care. Very rarely birds have adopted similar behaviour with different objects, including lighted cigarette butts!

Smoke-bathing

Rooks, Jackdaws and Starlings have been observed 'bathing' in smoke. They sit on a smoking chimney and allow the smoke to pass through their feathers.

It is believed that this may be a version of anting behaviour, but using a different stimulus. It is tempting to think these actions help to remove mites and lice in the feathers, but as yet there is no proof of this.

A feature of late summer is migrating flocks of Swallows and martins sometimes landing on roofs and other surfaces facing the sun and engaging in a short period of sunning.

Flight

Of all birds' habits, it must be their flight that impresses humans most. We admire the precision of diving Gannets and the effortless soaring of raptors. We marvel at the formations of Starling flocks that make a kaleidoscope of patterns across the sky, and we mourn the loss of the Skylark's songflight from much of our countryside. If we spare a moment from our busy lives we can become mesmerised by the flight of a Swallow as it collects flies – swooping, twisting and turning in a seemingly endless search, but always perfectly coordinated and in total control of its flight.

TYPES OF FLIGHT

As we have already seen, birds that migrate have two main modes of flight: flapping and gliding. These are straightforward methods for covering distance in the air, but other modes of flight exist for more specific needs.

Fulmars ride the updraughts around the cliffs where they nest.

The Great Shearwater uses the currents of air close to the waves to help it on its great journeys across oceans. This flying method saves energy and appears quite effortless, even in stormy conditions.

Gliding flight

Large, broad-winged species such as eagles will glide and soar on their migrations when they can, using air currents to gain lift and then gliding to save energy. They will also use flapping flight at other times; when moving around their territory and home range or when carrying heavy prey. At other times they may soar over their territory, which is an energy-efficient way of displaying or hunting.

Another form of soaring is known as 'dynamic soaring' and can be seen at sea when shearwaters and albatrosses, and sometimes Gannets, use the air currents that develop between the waves. Thus seabirds can travel apparently effortlessly by using the cushion of air above the waves and then rising, banking and falling again as they use the turbulence to help them on their travels. This allows them to fly efficiently in conditions that on the face of it look impossible.

Most seabirds are masters of this flight over the waves, but the albatrosses, being the largest, are perhaps the most spectacular, and use this form of travel to fly up to 600 miles in a day.

A male Great Spotted Woodpecker pushes air down with its broad wings as it takes off in an uncharacteristic pose. In level flight it switchbacks, almost closing its wing between flaps.

Red Kites are masters of flight as they use the air currents and updraughts and are capable of appearing to 'hang on the wind'. Their long wings with fingered tips and long tails allow the birds to steer and change course with minimum effort.

Flapping flight

Flapping flight requires a bird to push the air below it downwards with sufficient momentum to match its own body weight. A slow-flying bird will need to flap harder and use more energy than a fast-flying individual, provided the faster bird does not cause too much drag.

Different species and groups of birds have different and characteristic patterns of flapping flight. Small finches such as Goldfinch seem to bounce along, with wings momentarily closing. Woodpeckers also close their wings for longer in flight, and their flight pattern typically zigzags. Crows, especially Jays, appear to row themselves across open spaces, while screaming parties of Swifts in late summer have flickering wings and sometimes appear to almost turn over in flight. Common Sandpipers also have flickering wing-beats on bowed wings.

Songflights and hovering

Hovering requires a bird to head into the wind and fly, so that its flapping wings make enough speed to match the wind, and therefore remain stationery. Uplift from the wind can sometimes allow birds to 'hang' on the spot without flapping for short spells.

Skylarks' songflights normally last between two and four minutes, although can continue much longer, even up to 30 minutes. All this time they are above their territory flying into the wind with wings beating about seven to 10 times a second – just enough to stay more or less still. If the wind is blowing at 6m per second the bird will be stationary with virtually no additional effort needed.

Skylarks are specially adapted for their songflights, with relatively long and broad wings – especially the males. It has been demonstrated that those males with the largest wing areas (longer and broader) have longer songflights and are preferred by females.

Other birds hover when hunting and use wind or updraughts. Red Kites and Buzzards 'hang in the wind' with wings hardly moving. Kingfishers and terns will also hover when fishing. Again they use the wind to help them remain stationary in the air, but their hovering usually requires considerable flapping.

Kestrels require some wind in order to be able to hover and hunt successfully. Although strong winds may buffet a hovering Kestrel, and its wings will beat vigorously while its body gets blown about, the bird's head will remain perfectly still as it searches the ground for its prey.

During early speed trials for naval vessels in Scotland, a ship travelling at almost 37mph into the wind was frequently being overtaken by Puffins, Razorbills and Guillemots!

Kestrels are well known for their ability to hover, although they frequently hunt from a post or tree – especially when there is a lack of wind. When hovering their bodies are angled at about 45 degrees and the movements of their wings depend on the conditions and strength of the wind. Sometimes they are almost motionless as little wing flickers match the force of the wind, but in blustery conditions the position will be controlled by more vigorous flapping and tail movements.

In North and South America there are many species of hummingbirds, which are adapted to be able to hover in front of flowers, so that their bills and especially their long tongues can probe into exotic blooms to reach the nectar that they need to provide their energy. Their wings are modified so that they move freely in all directions, which allows for an almost rotary movement, more like an insect than a bird, and they beat their wings at 22–78 times a second.

HOW FAST CAN BIRDS FLY?

We humans are always fascinated by birds' flight speed. Sometimes we get an impression of ground speed when a bird happens to fly alongside a car or train. It is not unusual, for example, to find a Woodpigeon is flying at 20–30mph.

In general, large birds fly faster than small ones. It has been estimated that small songbirds can probably fly at up to 14–18mph, larger birds up to 27mph and ducks and large waders at up to 45mph. There is even a report of a Cormorant flying at 58mph.

In migratory flights birds tend to fly higher, especially over the sea. At a greater height they benefit from reduced air density that creates faster air speed, and may have the advantage of following winds. Recent data from migratory birds has provided further evidence of their amazing flights. Migrating Chaffinches fly at 18mph, Swallows at 27mph, Skylarks at 33mph and Bewick's Swans at 40mph. These species all fly with a flapping flight. As a comparison, Honey Buzzard, a typical large soaring species, travelled at 24mph

Knots have been recorded travelling at 35mph on their migrations between Australia and China, and Bar-tailed Godwits that travel from New Zealand to Alaska on an apparently non-stop flight covered this distance in seven days, achieving an average speed of 39mph.

One species that has often intrigued people by its power and speed of flight is the Peregrine. In level flight it is thought to average around 60mph, but when stooping headlong on its prey, it achieves considerably greater speed.

Claims of over 100mph have long been made for this short dramatic stoop, but accurate measurement in such a short time frame is difficult. The current best measurement is

Brent Geese migrating from Alaska to Mexico achieved average speeds, in different years, of between 35 and 50mph.

170mph for a stoop at 30 degrees and 220mph for a stoop at 45 degrees. However, mathematical calculations make the speed more likely to be 120–150mph. Whichever is true, this is still a very impressive example of controlled flight, and is the fastest known bird 'flight' (albeit gravity-assisted).

WEIGHT RESTRICTIONS?

Feathers have evolved from reptilian scales, and the flight feathers provide a large but very lightweight surface area as well as the strength and flexibility needed for flight. The avian skeleton also weighs very little compared to that of other vertebrates.

Birds' bones have to be strong to withstand all the normal forces of flying, but they need to be lightweight as well. Many bird bones are not only light but hollow, and reinforced internally with struts or filaments of bone that provide strength without compromising weight.

The ability to lay eggs, and therefore allow the young to grow outside the body, enables female birds to retain their flight ability at all times, and was inherited from their reptilian ancestors.

As we have already seen when considering migration, birds are capable of carrying increased weight in the form of body fat when they fly, and many small songbirds almost double their body weight before setting out. While this fat provides fuel for the journey it is additional bulk to carry, especially when making the climb to a suitable altitude for migration.

FORMATION FLYING

Geese are masters of the art of formation flying and are well known for their 'V' shaped flight patterns. These flocks allow individual birds to benefit from the actions of the bird in front by slipstreaming, and getting reduced drag and greater lift. It has been estimated that there is an energy saving of 15–20% for the followers. Oldest males are usually in the front of their family groups, but the position of the family within a large flock may change from time to time.

The speed of the **Peregrine** has long been debated. In its stoop on its prey it achieves speeds of around 150mph and probably sometimes faster than that.

HOW DO THEY DO IT?

Birds differ from bats, and extinct flying reptiles (pterosaurs), by having their wings independent from their legs. Birds' wings are connected to their bodies at one point only, and are powered by the large pectoral muscles.

Wing shape varies from short and rounded in the Wren to long and thin in the Manx Shearwater. The shape is adapted to the bird's lifestyle; its habitat and its feeding habits. Wing size is often described by scientists in terms of its wing-loading, which is the body weight divided by the wing area. The wing shape is described as the aspect ratio – the wingspan squared divided by wing area.

Birds with a low aspect ratio, such as many small songbirds, tend to have short round wings and their low wing loading allows them to take on extra fat for longer flights. Those with short wings but a high aspect ratio, such as the auks, have thinner wings and higher speeds with less power. Birds with long wings and a high aspect ratio are aerodynamically efficient, like the albatrosses, provided they do not increase their body weight unduly.

The membrane that was used for powered flight by pterosaurs was attached to their 'arms' and legs, and a similar arrangement can be seen today in bats.

LARGEST FLYING BIRDS

Birds of the same species vary in size and often there is a size difference between male and female. When it comes to flight, both weight and overall size are limiting factors.

The heaviest flying bird is the Great Bustard, males of which may reach 16kg, and then there are the swans – with Whooper weighing in at up to 14kg, making it one of the largest migratory birds in Europe.

The Andean Condor of South America, with huge wings spanning 2.8–3.2m, is often claimed as the largest flying bird because of its huge wing area. However, the Wandering Albatross, which has very narrow wings compared to the condor, may have a longer wingspan (wing-tip to wing-tip) of up 3.7m.

EXPERT AND AGILE

The prowess of bird flight never ceases to amaze. Even relatively familiar behaviour is remarkable if you find time to stop and watch. Returning to the Swallows that started this chapter, the juveniles, just out of their nest, will often sit on a wire where they will be fed by their parents. The adult will not land but hover briefly in front of a youngster and push some food into its gaping mouth. The dexterity and precision of this manoeuvre in flight is quite remarkable.

Even more remarkable is seeing Swallows feeding flying young while both are in flight and, again, the mastery of flight and coordination that is required is breath-taking.

We have already seen how songflights and aerial displays can be used to signal territorial ownership, but some of these displays can also demonstrate aerobatic skill.

The tumbling display of a Lapwing as it drops, almost to ground level, and then zigzags over its mate before banking steeply and going round again is one of the evocative sights of spring. The talon-grappling of White-tailed Eagles, as one flies under the other and rolls over in flight, seems an impossible manoeuvre for such a large bird. And then there are the Rooks. Ubiquitous birds of village and countryside, in their autumn

The Whooper Swan is one of the heaviest birds to be able to fly, yet many make a regular migration from Iceland to western Europe each year.

flocks they will suddenly stall and tumble and then pull out of an untidy dive – it looks great fun, as if the birds really enjoy being masters of the air.

STARTING YOUNG

While young Swallows take food in flight soon after leaving the nest, there are other examples of young birds taking to the air even earlier, although with less grace. Grey Partridges, being ground nesters, are highly vulnerable to many predators, and their wing feathers develop more quickly than their other plumage. They are able to take to the air when only a fraction of the size of their parents – at 15 days – whereas they are not fully grown until almost 100 days.

FLYING UNDER WATER

Obviously underwater flying is not real flight, but it has many of the same wing movements required as flying in the air. The birds that dive to find food mostly hunt fish, but sometimes they search for other food such as pondweed and freshwater mussels. Those that feed on fish need to be fast-moving and adroit if they are going to be successful, and in some species the wings are essential for propulsion as are the webbed feet.

It requires extreme accuracy for an adult Swallow to feed its chicks while flying – yet this is quite normal behaviour during the first few days after the young have left the nest.

Displaying Lapwings rapidly twist and turn and their flight feathers 'hum' during their exciting displays.

Guillemots and other members of the auk family use their wings to 'fly' under water as they chase small fish.

Most do not stay under water for long – seldom more than a minute or two – during which time they chase and capture their prey.

The streamlining that helps in aerial flight also reduces drag when swimming under water, and the short stiff wings of a bird such as a Puffin make ideal paddles. These adaptations have been taken to extremes in the case of the penguins.

In the northern hemisphere the diving seabirds we know as the auks (Puffin, Razorbill, Guillemot, etc.) have many of the characteristics of the penguins of the southern hemisphere. The families are not related, but have followed a similar line of evolution (known as convergent evolution). Auks, however, are not flightless, with the exception of the Great Auk. This trait led to its downfall – it was hunted to extinction in 1844.

FLIGHTLESSNESS

Penguins are not the only birds that are flightless, although

Take-off trouble

Gannets are great fliers and competent swimmers, but from the water they find it hard to get airborne, unless the wind over the water is quite strong. A bird in calm weather, especially if heavy after fishing, will attempt to run along the surface in order to take off. Gannets' wings may be aerodynamic for soaring and gliding, but the high wing-loading does not allow them to lift off from the calm water without losing valuable energy.

all flightless species descended from ancestors that were undoubtedly fully flighted. Usually they are birds of remote places where the ability to fly ceased to have any advantage. The penguins of Antarctica and the southern continents are masters of the seas and travel huge distances by swimming.

The Flightless Cormorant of Galapagos is safe on its remote, predator-free islands. The Short-winged Grebe of Lake Titicaca high in the Andes is also flightless, but lives on deep waters in a remote wilderness. The Giant Coot of South America becomes flightless when adult, though young birds can fly. All in all there are about 50 species alive today that are virtually or completely flightless. Many others, sadly, have become extinct, often as a result of travelling humans introducing non-native predators such as cats and rats to their habitats.

More familiar birds may also be flightless, but only for a time. Ducks like Mallard are flightless for up to a month when moulting and so are many other water birds. Guillemots and Razorbills are also flightless when moulting, but this coincides with the time they are living at sea.

Three species of flightless Kiwis nest in New Zealand. This is the Brown or Common Kiwi. It is mainly nocturnal feeding, on worms and other invertebrates which it locates by smell.

Vision and other senses

There are few more enchanting sights than a hunting Barn Owl. Often glimpsed at dusk, but sometimes in full daylight, the bird will fly silently along a verge or across a field, its senses alert as it examines the ground below. Then, with wings raised and legs dangling, it drops from view, perhaps onto an unsuspecting vole or shrew. These are moments to treasure – our chance to glimpse a world that is usually hidden by darkness.

But what are the senses that allow an owl to hunt, not just at dusk but also in the dead of night, and what other senses can birds call up to help their survival?

EYES

Birds have very well-developed and efficient eyesight – they need to as their life depends on it, for finding food, avoiding danger and selecting a mate. Their eyes are also surprisingly large – sometimes each eye is as large as their brain, and the eyes of larger birds of prey and owls can be the size of human eyes.

Barn Owls are able to locate their prey in total darkness.

The shape of birds' eyes, and therefore their efficiency, varies

but they probably see objects two or three times more sharply than a human's eyes. It has been calculated that some raptors can see small prey from up to a mile away.

The eyes are relatively immobile although some birds can move them a little, especially those like the Cormorant that hunt under water. In general birds compensate for their lack of eye movement by having flexible necks. A Blackbird will tilt its head on one side to look for movements on the ground, and an owl can turn its head through 180 degrees and so can see behind without having to move its body.

Owls have various adaptations to help them locate their prey in the dark. Their large eyes have a short focal length, which results in very little light being lost. The owl retina also has a large number of light-sensitive cells (rods) which are not good at transmitting colour or detail to the brain, but makes the eye very sensitive to light, even at low intensity. Powerful muscles within the eye help the lens to focus very quickly, and the forward-facing position of owl eyes provides binocular vision – meaning that the bird can judge distance very accurately.

Like us, birds have a fovea – a concentration of colour-sensitive cone cells on their retinas near the optical nerve, corresponding to the centre of their visual field – and the fovea is more highly developed in those species that mainly hunt by sight. Some specialist hunters, such as terns, kingfishers and hummingbirds, have a second fovea that is thought to improve their binocular vision and therefore helps them judge critical distances.

There is an old story that if you find a perched owl it will follow your movement by turning its head, and if you walk round its perch three times its head will fall off! Although obviously not true, this shows that for a long time owls' flexible necks have been noted as being exceptional among birds. In fact they can turn through 180°.

Protecting eyes

All birds have a third eyelid, properly called the nictitating membrane, which moves quickly and is seldom seen. It is normally transparent and wipes the eye. In owls it is opaque and is drawn across the eye as the bird strikes its prey, or when it feeds its young or when preening. This means that at these times its vision is obscured. To compensate for the lost vision it is thought the bristles around the bill are sensitive, in the way that a mammal's whiskers are sensitive.

Ultraviolet and polarised light

It is apparent that some bird species see a wider range of the wavelength spectrum than humans, including ultraviolet light. This helps a Kestrel hunting small mammals (see page 163), and it also helps hummingbirds that feed on nectar from flowers that have UV-specific colours and patterns.

It is now believed that birds, especially those that migrate, can recognise and use polarised light, which provides an additional navigational tool if the sun is obscured.

GOOD HEARING

Given the importance of bird songs and calls it is no surprise that birds have acute hearing as well. Owls need this faculty even more when hunting in the dark, when they use their ears to locate their prey – indeed it has been shown experimentally that a Barn Owl is able to hunt successfully by locating its prey in total darkness.

Night-hunting owls' ears are positioned asymmetrically, with one higher on the skull than the other, which means sound waves coming from directly above or below the bird strike the ear drums at slightly different times. This helps the owl pin-point its prey – it is an impressive location system that

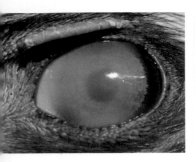

A semi-transparent nictitating membrane can be drawn across the eye to protect it and clean it without total loss of vision from blinking with a dull eyelid.

A sense of taste

Birds do have a sense of taste but this is not highly developed, with birds having only 25–70 taste buds whereas humans have 10,000. There may be enough information derived from this lesser sense to help a bird select its food, but it appears not to be of major use.

is further helped by the facial discs of specially adapted and rather stiff feathers that reflect sound into the bird's ears.

In addition owls, especially Barn Owl, have long wings and a low wing-loading, meaning they can fly slowly, and their feathers are covered with a velvet-like pile that further helps to deaden sound as they fly – helping them to hear even better as well as reducing the downdraught that might warn prey of their approach.

The Oilbird, a nightjar-like nocturnal species from South America and the Caribbean, is one of a handful of birds that can use echolocation to navigate. It makes clicking sounds, and listens for the subsequent echoes to build a mental picture of its surroundings.

HOW DO THEY SMELL?

There has been little research into how much birds can smell. Evidence indicates the ability to smell is well developed, but that for most species it is not important in their normal lives. However, American vultures, such as the Turkey Vulture, appear to use smell to locate carcasses. We also know that some seabirds, especially petrels, have a well developed olfactory lobe, the part of the brain used in smell.

Colonies of birds such as Manx Shearwater would be judged 'smelly' by most humans, but how important this is to the birds and their ability to find their burrows is uncertain. Experiments indicate smell is not used by Manx Shearwaters, but in the South Atlantic studies of prions (petrel-like birds) and Leach's Petrels have indicated that smell is used by these birds to locate their nest-sites.

It seems more certain that petrels use their sense of smell to find food. Birdwatching trips which throw out 'chum', a smelly concoction of fish offal, quickly attracts a selection of seabirds, including the petrels and shearwaters, and experiments in Canada show that these birds tended to approach smelly bait from down wind, strongly suggesting that they were smelling rather than seeing this food or finding it from the ocean currents.

The facial disc of a Short-eared Owl is made up of special feathers that reflect sounds into the bird's ears very efficiently.

Acknowledgements

I bought my first ornithological reference book when I was about 10 years old and I have collected books and papers on the subject ever since then. These days I also have a growing index of web references as well. In short, it is difficult to acknowledge all sources of information contained between these pages.

Some books are absolute 'bibles' when it comes to basic information, among them titles such as *A Dictionary of Birds* (Campbell and Lack, Poyser, 1985) and *Birds of the Western Palearctic* (Cramp, Simmons and Perrins, OUP, 1977–1994). *The Migration Ecology of Birds* (Newton, Academic Press, 2009) is also a fundamental work of reference on one important aspect of bird behaviour.

Two series of books helped provide much of the detailed information: the *New Naturalist* books (published by Collins) and the monographs and related titles published by T & A D Poyser. Especially useful were *Tits* (Perrins, Collins, 1979), *Finches* (Newton, Collins, 1972), *The Golden Eagle* (Watson, Poyser, 1997) and *The Peregrine* (Ratcliffe, Poyser, 1980). In addition, the books by David Lack, especially *The Life of the Robin* (Witherby, 1943) and *Swifts in a Tower* (Methuen 1956) were monumental works of their time that set a pattern from which we benefit today.

Birds Britannica (Cocker and Mabey, Chatto & Windus, 2005) is a treasure-house of stories, and so are the short notes published in the monthly journal *British Birds*.

The British Trust for Ornithology's publications (*Bird Study* and *BTO News*) carry reports of recent research and the website has a wonderful collection of data, including longevity data of British birds based on ringing recoveries. Of course, this data would not exist if it was not for the vast number of hours of recording and observation by huge numbers of enthusiasts over many years.

Many photographers have made available wonderful images which bring this fascinating subject to life in a way my words can never quite achieve – please enjoy both the beauty of the subject and the technical excellence of the photographers.

Last but not least, I am grateful to people who worked with me on the project. Nigel Redman understood what I was trying to achieve and commissioned the work. Lisa Thomas gave great encouragement and helped devise a sensible plan, and later Julie Bailey took over the reins and pushed the production ahead in a truly professional way. Finally, Marianne Taylor had the unenviable task of editing my first draft and also brought her own extensive knowledge to help shape the final content, which Nicki Liddiard at Nimbus Design was able to use to create these delightful page designs.

Peter Holden

Index

Image credits

Bloomsbury Publishing would like to thank the following for providing photographs and for permission to reproduce copyright material. While every effort has been made to trace and acknowledge all copyright holders, we would like to apologise for any errors or omissions, and invite readers to inform us so that corrections can be made in any future editions of the book.

1 top, David Tipling/davidtipling.com; 1 bottom, David Tipling/davidtipling.com; 2 bottom, David Tipling/davidtipling.com; 6, Tui De Roy/Minden Pictures/FLPA; 8 top, David Tipling/davidtipling.com; 9, Popperfoto/Getty Images; 10, David Tipling/davidtipling.com; 11, Popperfoto/Getty Images; 12, Paul Sawer/FLPA; 14, Giacomo Jacquerio/Getty Images; 16–17, David Tipling/davidtipling.com; 18, Frans Lanting/FLPA; 19, Scott Leslie/Minden Pictures/FLPA; 20, John Hawkins/FLPA; 21, Tony Hamblin/FLPA; 22, Vishnevokdy Vasily/Shutterstock; 23, R. Peterkin/Shutterstock; 24, Steve Gettle/Minden Pictures/FLPA; 25, John Hawkins/FLPA; 26, Adri Hoogendijk/FN/Minden/FLPA; 27, Attila Steiner; 28, David Hosking/FLPA; 20, Mike Lane/FLPA; 30, Michael Durham/FLPA; 31, Jos Korenromp/FN/Minden/FLPA; 32 main, ImageBroker/Imagebroker/FLPA; 32 inset, Krystyna Szulecka/FLPA; 33 top, Martin H Smith/FLPA; 33 bottom, Andrew Parkinson/FLPA; 34, Justus de Cuveland/Imagebroker/FLPA; 35, Paul Hobson/FLPA; 37 top, Roger Wilmshurst/FLPA; 37 bottom, T. S. Zylva/FLPA; 38, Tui De Roy/Minden Pictures/FLPA; 39, Derek Middleton/FLPA; 41, Michael Callan/FLPA; 42, John Eveson/FLPA; 43, David Tipling/davidtipling.com; 44, Mark Sisson/FLPA; 45, Jiri Foltyn/Shutterstock; 46 top, Alanfoto/Shutterstock; 46 bottom, Roger Wilmshurst/FLPA; 47, Winfried Wisniewski/Minden Pictures/FLPA; 48, Ed Phillips/Shutterstock; 50, John Watkins/FLPA; 51, David Tipling/FLPA; 52, Hans Menop/FN/Minden/FLPA; 53 top, Roger Wilmshurst/FLPA; 53 bottom, Danny Ellinger/Foto Natura/Minden Pictures/FLPA; 54, Steve Young/FLPA; 55, Andrew Bailey/FLPA; 56, Roger Wilmshurst/FLPA; 58, Peter Cairns/Foto Natura/MINDEN PICTURES; 59 top, Paul Sawer/FLPA; 59 bottom, Konrad Wothe/Minden Pictures/FLPA; 60, David Tipling/FLPA; 61, ImageBroker/Imagebroker/FLPA; 62, Tim Graham/Getty Images; 63, David Tipling/FLPA; 64 left, Wayne Hutchinson/FLPA; 64 right, John Hawkins/FLPA; 65, David Hosking/FLPA; 66, Robin Chittenden/FLPA; 67, Bill Baston/AGAMI; 68, David Tipling/davidtipling.com; 69, Markus Varesvuo; 70, David Tipling/davidtipling.com; 71, David Tipling/davidtipling.com; 72, Mark Sisson/FLPA; 73, Frans Lanting/FLPA; 74, John Hawkins/FLPA; 75, Keith Brockie; 76, David Tipling/davidtipling.com; 77, David Tipling/davidtipling.com; 78, Bence Mate/AGAMI; 79, David Tipling/davidtipling.com; 80, Silvestris Fotoservice/FLPA; 81, David Tipling/davidtipling.com; 82, Roger Tidman/FLPA; 83, Markus Varesvuo; 84, David Tipling/davidtipling.com; 85, Erica Olsen/FLPA; 86, David Tipling/FLPA; 87, Reinhard Hölzl/Imagebroker/FLPA; 88, Simon Litten/FLPA; 89, Marianne Taylor; 90, Hugh Clark/FLPA; 91, Dickie Duckett/FLPA; 92, David Tipling/davidtipling.com; 93, Paul Hobson/FLPA; 94, Sean Hunter/FLPA; 95, ImageBroker/Imagebroker/FLPA; 97, Markus Varesvuo; 98 top, Andrew Parkinson/FLPA; 98 bottom, Marianne Taylor; 99, Gary K Smith/FLPA; 100 top, Derek Middleton/FLPA; 100 bottom, Yva Momatiuk & John Eastcott/Minden Pictures/FLPA; 101, Wendy Dennis/FLPA; 103 top, Daniele Occhiato/AGAMI; 104 top, Paul Sawer/FLPA; 104 bottom, David Hosking/FLPA; 105, Malcolm Schuyl/FLPA; 106, Malcolm Schuyl/FLPA; 107, Marcel van Kammen/Minden Pictures/FLPA; 108, David Tipling/FLPA; 109, John Watkins/FLPA; 110 top, Tony Hamblin/FLPA; 110 bottom, Steve Young/FLPA; 111 top, Robin Chittenden/FLPA; 111 middle, Terry Whittaker/FLPA; 111 bottom, Joe Gough/Shutterstock; 112, Erica Olsen/FLPA; 113, Roger Tidman/FLPA; 114, Tony Hamblin/FLPA; 115, Marianne Taylor; 116, David Tipling/FLPA; 117, David Tipling/davidtipling.com; 118, Marianne Taylor; 119, Michael Durham/FLPA; 120, Gary K. Smith/FLPA; 121, David Tipling/davidtipling.com; 122, Roy de Haas/AGAMI; 123, Roger Tidman/FLPA; 124, Flip De Nooyer/FN/Minden/FLPA; 125, IMAGEBROKER, MICHAEL KRABS,Imag/Imagebroker/FLPA; 126 top, Markus Varesvuo; 126 bottom, David Tipling/FLPA; 127, Krystyna Szulecka/FLPA; 128, Harri Taavetti/FLPA;